ENERGY POLICIES, POLITICS AND PRICES

U.S. ENERGY PRODUCTION AND CONSUMPTION

MAJOR FACTORS, INFLUENCES, AND TRENDS

ENERGY POLICIES, POLITICS AND PRICES

Additional books in this series can be found on Nova's website under the Series tab.

Additional e-books in this series can be found on Nova's website under the e-book tab.

ENERGY POLICIES, POLITICS AND PRICES

U.S. ENERGY PRODUCTION AND CONSUMPTION

MAJOR FACTORS, INFLUENCES, AND TRENDS

KENDRA FLETCHER
EDITOR

nova publishers
New York

Library of Congress Cataloging-in-Publication Data

ISBN: 978-1-63482-182-7

Published by Nova Science Publishers, Inc. † New York

CONTENTS

PREFACE

This book provides information on U.S. production and consumption of fossil, nuclear, and renewable energy from 2000 through 2013 and major factors, including federal activities, that influenced energy production and consumption levels. It also provides information on other federal activities that may have influenced aspects of U.S. energy production and consumption from 2000 through 2013 but were not targeted at a specific energy source, as well as information on federal research and development.

Chapter 1 - Federal energy policy since the 1970s has focused primarily on ensuring a secure supply of energy while protecting the environment. The federal government supports and intervenes in U.S. energy production and consumption in various ways, such as providing tax incentives, grants, and other support to promote domestic production of energy, as well as setting standards and requirements.

GAO was asked to provide information on federal activities and their influence on U.S. energy production and consumption over the past decade. This report provides information on U.S. production and consumption of fossil, nuclear, and renewable energy from 2000 through 2013 and major factors, including federal activities, that influenced energy production and consumption levels. It also provides information on other federal activities that may have influenced aspects of U.S. energy production and consumption from 2000 through 2013 but were not targeted at a specific energy source, as well as information on federal research and development.

GAO analyzed DOE historical data on energy production and consumption, reviewed studies and reports from federal agencies and governmental organizations on federal energy-related activities, and analyzed data on federal spending programs and tax incentives, among other things.

GAO is not making recommendations in this report. DOE, the Department of the Treasury, and the U.S. Department of Agriculture reviewed a draft of this report and provided technical comments that GAO incorporated as appropriate.

Chapter 2 - Energy policy in the United States has focused on three major goals: assuring a secure supply of energy, keeping energy costs low, and protecting the environment. In pursuit of those goals, government programs have been developed to improve the efficiency with which energy is utilized, to promote the domestic production of conventional energy sources, and to develop new energy sources, particularly renewable sources.

Implementing these programs has been controversial because of varying importance given to different aspects of energy policy. For some, dependence on imports of foreign oil, particularly from the Persian Gulf, is the primary concern; for others, the indiscriminate use of fossil fuels, whatever their origin, is most important. The contribution of burning fossil fuels to global climate change is particularly controversial. Another dichotomy is between those who see government intervention as a positive force and those who view it as a necessary evil at best.

In: U.S. Energy Production and Consumption
Editor: Kendra Fletcher
ISBN: 978-1-63482-182-7

Chapter 1

ENERGY POLICY: INFORMATION ON FEDERAL AND OTHER FACTORS INFLUENCING U.S. ENERGY PRODUCTION AND CONSUMPTION FROM 2000 THROUGH 2013*

United States Government Accountability Office

WHY GAO DID THIS STUDY

Federal energy policy since the 1970s has focused primarily on ensuring a secure supply of energy while protecting the environment. The federal government supports and intervenes in U.S. energy production and consumption in various ways, such as providing tax incentives, grants, and other support to promote domestic production of energy, as well as setting standards and requirements.

GAO was asked to provide information on federal activities and their influence on U.S. energy production and consumption over the past decade. This report provides information on U.S. production and consumption of fossil, nuclear, and renewable energy from 2000 through 2013 and major factors, including federal activities, that influenced energy production and

* This is an edited, reformatted and augmented version of a United States Government Accountability Office publication, No. GAO-14-836, dated September 2014.

consumption levels. It also provides information on other federal activities that may have influenced aspects of U.S. energy production and consumption from 2000 through 2013 but were not targeted at a specific energy source, as well as information on federal research and development.

GAO analyzed DOE historical data on energy production and consumption, reviewed studies and reports from federal agencies and governmental organizations on federal energy-related activities, and analyzed data on federal spending programs and tax incentives, among other things.

GAO is not making recommendations in this report. DOE, the Department of the Treasury, and the U.S. Department of Agriculture reviewed a draft of this report and provided technical comments that GAO incorporated as appropriate.

What GAO Found

According to the studies and reports GAO reviewed, several major factors, including federal activities, influenced U.S. production and consumption of fossil, nuclear, and renewable energy from 2000 through 2013. Examples of these factors include the following:

- **Fossil energy.** Advances in drilling technologies enabled economic production of natural gas and crude oil from shale and similar geological formations. These advances led to increases in domestic production of natural gas and crude oil beginning around 2008 and contributed to declines in domestic prices of natural gas, as well as lower prices for crude oil in some regions of the United States. Some federal activities also may have influenced these trends. For example, the federal government limited oil producers' liability associated with some oil spills, lowering the producers' costs for liability insurance. In addition, the federal government provided tax incentives encouraging production for oil and gas producers, resulting in billions of dollars in estimated federal revenue losses. Moreover, partly because of lower natural gas prices, domestic coal production decreased in recent years as utilities switched from coal to natural gas for electricity generation.
- **Nuclear energy.** Declining prices for a competing energy source—natural gas—may have led to decreases in the production and consumption of nuclear energy in recent years. Federal activities may

have also influenced this trend. For example, the Department of Energy (DOE) announced plans to terminate its work to license a disposal facility for certain nuclear power plant waste in 2009, creating uncertainty about how this waste would be managed. This uncertainty may have provided a disincentive for some nuclear power operators to stay in the market or expand capacity because of the cost of storing nuclear waste.

- **Renewable energy.** Federal tax credits for ethanol and federal policies requiring the use of ethanol in transportation fuels were major factors influencing an 8-fold increase in the production and consumption of ethanol from 2000 to 2013. In addition, state policies requiring the use of renewable energy in electricity production, as well as federal outlays and tax credits for renewable energy producers, were major factors influencing a 30-fold increase and a 19-fold increase in production and consumption of electricity from wind and solar energy, respectively, from 2000 to 2013.

According to the studies and reports GAO reviewed, other federal activities may have influenced aspects of U.S. energy production and consumption from 2000 through 2013 but were not targeted at a specific energy source. For example, the federal government strengthened energy efficiency standards for vehicle fuel economy and consumer products such as appliances and lighting, provided electricity and transmission services to customers through its power marketing administrations and the Tennessee Valley Authority, and spent billions of dollars helping low-income households cover heating and cooling costs. In addition, the federal government supported research and development targeting a wide range of energy-related technologies at government-owned laboratories and through funding to universities and other research entities.

ABBREVIATIONS

Btu	British thermal unit
CAFE	corporate average fuel economy
CBO	Congressional Budget Office
CRS	Congressional Research Service
DOE	Department of Energy
EIA	Energy Information Administration

EISA	Energy Independence and Security Act of 2007
EPA	Environmental Protection Agency
EPAct	Energy Policy Act of 2005
Interior	Department of the Interior
JCT	Joint Committee on Taxation
NRC	Nuclear Regulatory Commission
OMB	Office of Management and Budget
Recovery Act	American Recovery and Reinvestment Act of 2009
R&D	research and development
Treasury	Department of the Treasury
TVA	Tennessee Valley Authority
USDA	U.S. Department of Agriculture

* * *

September 30, 2014

The Honorable Lisa Murkowski
Ranking Member
Committee on Energy and Natural Resources
United States Senate

Dear Senator Murkowski:

Americans' daily lives, as well as the economic productivity of the United States, depend on energy. Most sectors of American society rely upon a readily available and affordable supply of energy to function. However, the production and consumption of energy involve economic and environmental trade-offs. For example, fossil energy provides relatively inexpensive sources of fuels that power automobiles, generate electricity, and power industrial processes. However, burning these fuels increases air pollution and greenhouse gas emissions, according to the Environmental Protection Agency (EPA). According to the Congressional Research Service (CRS), federal energy policy since the 1970s has focused primarily on ensuring a secure supply of energy while protecting the environment.[1]

In pursuing the goals of a secure energy supply and a healthy environment, the federal government subsidizes or otherwise provides support to energy companies and consumers.[2] For example, the federal government provides tax incentives and other support to promote the domestic production of energy (including the extraction of coal, oil, and natural gas) and the development of renewable energy (including wind and solar power). The federal government also intervenes in energy markets in other ways, such as setting standards and requirements (through laws and regulations) that may not be directed at specific sources of energy but that nonetheless may influence the types and quantities of energy that are produced or consumed. For example, the federal government promotes energy efficiency through appliance efficiency standards that are written into law and regulates pollutants that are created in the production and consumption of energy.

The federal government has established a number of important energy policies over the past decade, and several federal organizations have studied some of the costs associated with these policies. For example, Congress passed several key laws affecting energy producers and consumers over the past 10 years, including the Energy Policy Act of 2005 (EPAct), the Energy Independence and Security Act of 2007 (EISA), and the American Recovery and Reinvestment Act of 2009 (Recovery Act).[3] Several federal organizations, including the Department of Energy's (DOE) Energy Information Administration (EIA), CRS, and the Congressional Budget Office (CBO), have published reports identifying and quantifying aspects of federal support for energy production and consumption associated with these policies.[4] In general, these reports focused on costs associated with federal energy-related tax incentives, outlays, or loan or loan guarantee programs.

You asked us to provide a broader context for understanding how federal activities have influenced U.S. energy production and consumption over the past decade. In this report, we provide information on U.S. production and consumption of fossil, nuclear, and renewable energy from 2000 through 2013 and major factors, including federal activities, that influenced energy production and consumption levels. We also provide information on other federal activities that may have influenced aspects of U.S. energy production and consumption from 2000 through 2013 but were not targeted at a specific energy source, as well as information on federal support for research and development (R&D).

To provide information on U.S. production and consumption of fossil, nuclear, and renewable energy from 2000 through 2013, we reviewed and analyzed EIA historical data, as well as EIA articles and monthly and annual reports. We assessed the reliability of EIA's data and determined them to be sufficiently reliable for our purposes. To provide information on major factors, including federal activities, that influenced energy production and consumption levels, we reviewed our prior work. We also reviewed reports and studies from federal agencies and government organizations, including CBO, CRS, DOE, Department of the Treasury (Treasury), EIA, congressional Joint Committee on Taxation (JCT), Office of Management and Budget (OMB), and U.S. Department of Agriculture (USDA). To identify these reports and studies, we conducted searches of various databases, such as ProQuest and PolicyFile, and asked agency officials to recommend studies. In addition, we reviewed and analyzed data and documentation on outlays, royalties collected, excise taxes collected, tax expenditures, forgone royalties, and federal credit programs collected from DOE, Department of the Interior (Interior), JCT, OMB, and Treasury. Based on a review of this information, we assessed the reliability of these data and determined them to be sufficiently reliable for our purposes.

To provide information on other federal activities that may have influenced aspects of U.S. energy production and consumption from 2000 through 2013 but were not targeted at a specific energy source, and to provide information on federal support for R&D, we reviewed our prior work and the reports and studies from federal agencies and governmental organizations (as described above). We also reviewed and analyzed data and documentation on outlays, tax expenditures, and federal credit programs from DOE, JCT, OMB, and Treasury. Based on a review of this information, we assessed the reliability of these data and determined them to be sufficiently reliable for our purposes.

Overall, we relied on our prior work and the work of other federal agencies and governmental organizations to identify major factors that influenced U.S. energy production and consumption levels. However, this work may not have identified all of the relevant factors that influenced energy production and consumption from 2000 through 2013. In addition, we did not examine permitting issues on federal lands, including issues related to infrastructure and oil and gas development. Appendix I provides additional information on the scope and methodology of our review.

We conducted this performance audit from August 2013 to September 2014 in accordance with generally accepted government auditing standards. Those standards require that we plan and perform the audit to obtain sufficient, appropriate evidence to provide a reasonable basis for our findings and conclusions based on our audit objectives. We believe that the evidence obtained provides a reasonable basis for our findings and conclusions based on our audit objectives.

Background

U.S. Sources and Uses of Energy

The United States consumes energy from three major categories of sources: fossil, nuclear, and renewable. Fossil energy comes from coal, natural gas, crude oil, and petroleum products.[5] Nuclear energy comes from uranium, which is mined and processed into nuclear fuel. This fuel undergoes nuclear fission in a nuclear reactor to produce heat, which is converted into electricity using steam turbine technology. Renewable energy comes from a variety of sources, including biomass, which is organic material from plants and animals and includes liquid biofuels (such as ethanol and biodiesel), wood, and waste (such as municipal solid waste and agricultural byproducts); hydroelectric power; geothermal; wind; and solar.[6] Many sources of renewable energy are converted into electricity before being consumed. The United States domestically produces most of the energy it consumes. However, it also imports a portion, mostly in the form of crude oil and petroleum products.

In 2013, the United States consumed over 97 quadrillion British thermal units (Btu)[7] of energy, including over 12 quadrillion Btus of imported energy, according to EIA data.[8] As shown in figure 1, most of this energy (or about 82 percent) came from fossil energy sources. The rest came from renewable and nuclear energy sources.

There are four major sectors of the U.S. economy that consume energy at the point of end use:

- the industrial sector, which includes facilities and equipment used for manufacturing, agriculture, mining, and construction;
- the transportation sector, which generally comprises vehicles (such as cars, trucks, buses, trains, aircraft, and boats, among others) that transport people or goods;
- the residential sector, which consists of homes and apartments; and
- the commercial sector, which includes buildings such as offices, malls, stores, schools, hospitals, hotels, warehouses, restaurants, and places of worship, among others, as well as federal, state, and local facilities and equipment.

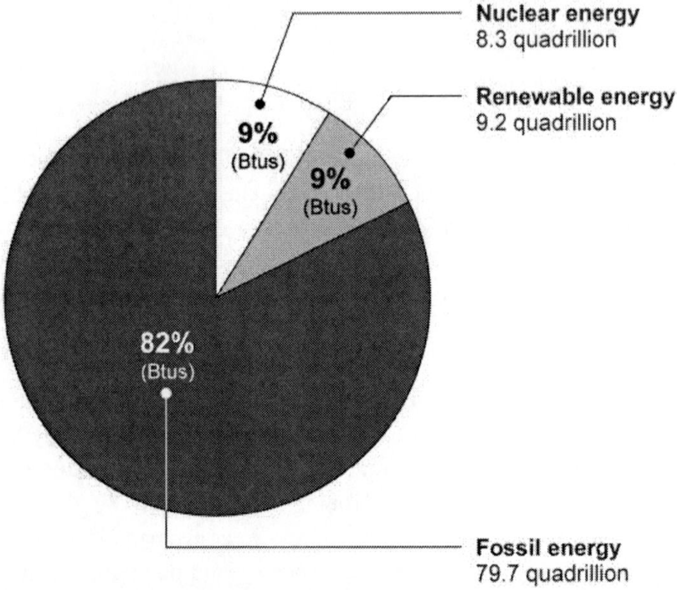

Source: GAO analysis of Energy Information Administration data. | GAO-14-836.
Note: Btu stands for British thermal unit.

Figure 1. U.S. Consumption of Energy by Source, 2013.

End-use sectors obtain energy from different combinations of sources. The industrial sector mainly consumes natural gas and electricity but also uses some petroleum products as feedstock. The transportation sector mainly consumes gasoline, diesel, and jet fuel; it also consumes biofuels and natural gas, as well as small amounts of electricity. The residential and commercial sectors mainly consume energy from electricity and natural gas but also use some petroleum products. As described above, every sector consumes electricity produced by the electric power sector, which takes electricity generated from fossil, nuclear, or renewable energy and delivers it to the end-use sectors through transmission and distribution lines. As shown in figure 2, the industrial sector consumed the largest share of energy (32 percent or 31.3 quadrillion Btus) in 2013, followed by the transportation, residential, and commercial sectors.

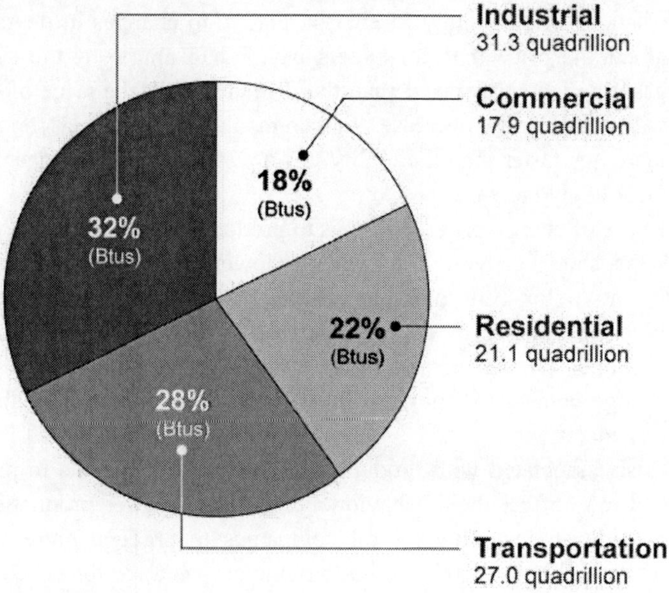

Source: GAO analysis of Energy Information Administration data. | GAO-14-836.
Note: Btu stands for British thermal unit.

Figure 2. U.S. Consumption of Energy by Sector, 2013.

Not all of the energy produced is available for consumption at the point of end use, mainly because energy losses occur whenever energy is converted from one form to another. For example, coal-fueled power plants produce electricity by burning coal in a boiler to heat water and produce steam. The steam, at tremendous pressure and temperature, flows into a turbine, which spins a generator to produce electricity. During this process, the burning of coal produces heat energy, some of which converts water into steam. In turn, some of the energy in the steam is converted into electrical energy. At each point in this process, some of the original energy contained in the coal is lost.[9] According to EIA, about two-thirds of the energy consumed to generate electricity is lost in conversion, and most of these losses occur in fossil-fueled and nuclear power plants that generate steam to turn turbines.[10]

According to general economic principles, a number of factors may affect the production and consumption of energy. These factors include the following:

- Changes in the supply of energy relative to changes in demand may affect the price that consumers pay. For example, if the supply of gasoline increases faster than the demand for it, the price of gasoline will most likely decrease. In contrast, if the demand for gasoline increases faster than the ability to supply it, the price of gasoline will most likely increase.
- Prices of energy provide signals to producers and consumers and may affect their behavior. For example, lower gasoline prices provide an incentive for consumers to consume more gasoline, while higher gasoline prices provide an incentive for consumers to consume less gasoline. (However, incentives or disincentives may not actually change behavior if they are insufficient to outweigh other factors or considerations.)
- Costs associated with producing energy provide signals to producers and may affect their behavior. For example, lower production costs provide an incentive for oil companies to produce more gasoline, while higher production costs provide an incentive for oil companies to produce less gasoline.

Factors Influencing Overall U.S. Energy Production and Consumption

Three important factors—energy efficiency, energy conservation, and the global economic recession of 2007 to 2009—had a major influence on U.S. energy production and consumption over the past decade but generally affected all energy producers and consumers rather than affecting only a specific source of energy. The first two factors are related but have distinct meanings. Energy efficiency is the use of technology that requires less energy to perform the same function, such as using a compact fluorescent light bulb in place of an incandescent bulb. Energy conservation is any behavior that results in the use of less energy, such as turning the lights off when you leave the room. Both energy efficiency and conservation reduce our consumption of energy and thereby influence the total amount of energy produced and consumed in the United States.

Some data suggest that the United States increased the efficiency with which it used energy from 2000 to 2013. For example, according to Federal Highway Administration data, the fuel economy of U.S. motor vehicles increased from an average of 16.9 miles per gallon in 2000 to 17.6 miles per gallon in 2012 (the latest year for which data are available). Another measure of the efficiency with which the United States uses its energy is to compare U.S. energy consumption with the U.S. gross domestic product, which is a measure of the value of all final goods and services produced within the United States in a given period.[11] By making this comparison, we can measure whether it takes less energy over time to produce the same value of goods and services. According to EIA data, the consumption of energy per dollar value of gross domestic product decreased from about 7,900 Btus per dollar in 2000 to about 6,200 Btus per dollar in 2013, indicating that the United States became more energy efficient.[12]

The third factor, the global economic recession of 2007 to 2009, resulted in reduced economic activity in general, which reduced energy demand and had long-term effects on global energy markets.[13] For example, according to a 2009 International Energy Agency report, the recession's effects on both energy producers and consumers included the following:[14]

- Energy companies drilled fewer oil and gas wells and reduced spending on refineries, pipelines, and power stations.
- Businesses and households spent less on energy-using appliances, equipment, and vehicles.
- Tighter credit and lower prices made investment in energy savings less attractive financially, while the economic crisis encouraged energy consumers to reduce overall spending. As a result, the deployment of more energy-efficient equipment was delayed.
- Equipment manufacturers were expected to reduce investment in research, development, and commercialization of more energy-efficient models.

Key Methods of Federal Support and Interventions

The federal government supports or intervenes in U.S. energy production and consumption through a number of key methods, including (1) setting standards and requirements, (2) directly providing goods and services, (3) assuming risk, (4) providing funds, and (5) collecting or forgoing revenue from taxes or fees. The federal government also conducts and provides funding for energy-related R&D.

Setting Standards and Requirements

Through laws and regulations, the federal government sets standards and requirements for (or prohibitions against) certain activities.[15] Some laws and regulations focus on affecting economic activity by controlling prices, output, or the entry and exit of firms in a market. For example, under the Atomic Energy Act, as amended,[16] the Nuclear Regulatory Commission is responsible for issuing licenses to commercial nuclear reactors and conducting oversight of activities under such licenses to protect the health and safety of the public, among other things. Without such licenses, firms cannot operate nuclear reactors and generate electricity from nuclear energy. Other laws and regulations focus on the effects of economic activity on the health and welfare of citizens. For example, under the Clean Air Act,[17] EPA is responsible for regulating emissions of a variety of air pollutants from coal-fueled power plants and other energy producers. While laws and regulations vary widely in how they are designed, they tend to function by influencing the decisions of producers and consumers in the market. In addition, laws and regulations may impose a variety of costs, such as costs on regulated entities to comply with the laws and regulations ("compliance costs"), and costs on government agencies to administer and enforce them. To the extent that the compliance costs affect the costs of engaging in particular energy-related activities, laws and regulations may change the behavior of energy producers and consumers.

Directly Providing Goods and Services

The federal government provides some goods or services directly—that is, through a government agency—rather than providing funds to another entity to provide these goods or services. For example, the federal government may produce and sell electricity generated at federally-owned facilities and produce reports and information on energy markets, among other things. Government provision of goods or services may be deemed necessary to address certain circumstances, such as economic inequalities among segments of the public or a need for a good or service considered unlikely to be met by the private sector. Such activities may affect energy producers and consumers in different ways. For example, production and sales of electricity generated at federally-owned facilities may involve energy sources and prices that differ from those of electricity produced and sold by private market participants.

Assuming Risk

The federal government assumes risk (and potential costs associated with risk) in a number of ways, such as

- making direct loans—disbursing funds to nonfederal borrowers under contracts requiring the repayment of such funds either with or without interest;
- guaranteeing loans—providing a guarantee, insurance, or other pledge regarding the payment of all or a part of the principal or interest on any debt obligation of a nonfederal borrower to a nonfederal lender;
- limiting liability; and
- providing or subsidizing insurance.

By assuming some or all of the costs associated with risks for certain energy activities, the government may make those activities relatively less expensive, thus providing an incentive to pursue those activities. For example, if the federal government assumes the risk of default on a loan to a manufacturer of turbines (that generate electricity from wind energy), nonfederal lenders may offer a lower interest rate to the manufacturer than they would in the absence of the federal guarantee. Lowering the costs of capital for developers could result in certain projects being financed that would otherwise not be built.

Providing Funds

The federal government directly provides (or outlays) funds for different purposes.[18] In some cases, an agency may provide funds in the form of a grant. For example, USDA may provide grants to help farmers, ranchers, and rural small businesses purchase and install renewable energy systems. In other cases, an agency may need to purchase goods or services. For example, federal agencies purchase energy for their buildings, as well as vehicles and fuel for these vehicles. To the extent that federal outlays lower the cost associated with a particular activity, federal outlays may lead to changes in the behavior of energy producers and consumers.

Collecting or Forgoing Revenues

The federal government collects revenues using different methods. One prominent method is through the tax system, which includes personal income taxes, corporate income taxes, and excise taxes based on the value of goods and services sold, among other types of taxes.[19] The primary purpose of the federal tax system is to collect the revenue needed to fund the operations of the federal government. Taxes may alter taxpayers' behavior by inducing them to shift resources from higher-taxed uses to lower-taxed uses in an effort to reduce tax liability.

The federal government also collects revenues associated with its management of federal lands. The federal government owns and manages roughly 30 percent of the nation's total surface area (or about 700 million acres onshore). It also has jurisdiction and control over the outer continental shelf, which includes about 1.8 billion acres of submerged lands in federal waters off the coast of Alaska, in the Gulf of Mexico, and off the Atlantic and Pacific coasts.[20] The federal government leases federal lands for the production of oil, gas, minerals such as coal, or other resources. In exchange, the government generally collects revenues, including payments in the form of rents and bonuses, which are required to secure and maintain a lease, and royalties, which are based on the value of the minerals that are extracted.

Table 1. Types of Tax Expenditures

Type	Description
Exclusion	Excludes income that would otherwise constitute part of a taxpayer's gross income.
Exemption	Reduces gross income for taxpayers because of their status or circumstances.
Deduction	Reduces gross income due to expenses taxpayers incur.
Credita	Reduces tax liability dollar-for-dollar. Additionally, some credits are refundable meaning that a credit in excess of tax liability results in a cash refund.
Deferral	Delays recognition of income or accelerates some deductions otherwise attributable to future years.
Preferential tax rate	Reduces tax rates on some forms of income.

Source: GAO. | GAO-14-836

Note: These tax expenditures are identified in the Congressional Budget and Impoundment Control Act of 1974.

[a]Certain business tax credits, including some energy credits, are claimed as part of a general business tax credit. The total credit can reduce current year tax liability but is not refundable; unused general business credits may be carried back one year and forward 20 years.

However, the federal government may choose to forgo certain revenues. Tax expenditures are tax provisions that are exceptions to the "normal structure" of individual and corporate income tax necessary to collect federal revenue.[21] These preferences can have the same effects as government spending programs; hence the name tax expenditures.[22] The Congressional Budget and Impoundment Control Act of 1974[23] identified six types of tax provisions that are considered tax expenditures when they are exceptions to the normal tax, as described in table 1. Tax expenditures may affect the behavior of energy producers and consumers by providing an incentive to engage in certain types of activities. For some tax expenditures, forgone revenues can be of the same magnitude or larger than related federal spending for some mission areas.[24]

In addition to forgoing tax revenues, the federal government may choose to forgo revenues associated with its leases of federal lands and waters. "Royalty relief" is a waiver or reduction of royalties that companies would otherwise be obligated to pay for their leases of federal lands or waters. For example, the Outer Continental Shelf Deep Water Royalty Relief Act of 1995[25] mandated royalty relief for oil and gas leases issued in the deep waters of the Gulf of Mexico from 1996 to 2000.

Research and Development

Energy-related R&D takes place across a spectrum of activities, including basic research, applied research, and demonstration.[26] Basic research includes efforts to explore and define scientific or engineering concepts or to investigate the nature of a subject without targeting any specific technology. Applied research includes efforts to develop new scientific or engineering knowledge to create new and improved technologies. Demonstration includes efforts to operate new or improved technologies to collect information on their performance and assess readiness for commercialization and deployment for widespread use.

The federal government plays a critical role in supporting energy-related R&D, which may involve conducting R&D at government-owned laboratories or funding another entity to conduct R&D. For example, as one of the largest research agencies in the federal government, DOE spends billions of dollars every year on R&D to support its diverse missions, including advancing scientific research and technology development and ensuring efficient and secure energy, among other things. However, because long time lags may occur between basic research activities and activities related to commercialization and deployment, it is often difficult to link government-funded R&D to specific effects on energy production, consumption, and prices in the future.

DOE's R&D covers a broad range of activities, and DOE program offices manage 17 national laboratories.[27] The following DOE program offices and laboratories primarily support energy-related R&D:

- The Office of Science oversees six national laboratories with research areas focusing on energy: Ames Laboratory in Iowa, Argonne National Laboratory in Illinois, Brookhaven National Laboratory in New York, Oak Ridge National Laboratory in Tennessee, Pacific Northwest National Laboratory in Washington, and Princeton Plasma Physics Laboratory in New Jersey. The Office of Science is the nation's single largest funding source for supporting research in energy sciences.
- The Office of Nuclear Energy oversees the Idaho National Laboratory in Idaho. The office's primary mission is to advance nuclear power as a resource capable of meeting the nation's energy, environmental, and national security needs by resolving technical, cost, safety, proliferation resistance, and security barriers.
- The Office of Fossil Energy oversees the National Energy Technology Laboratory in Pennsylvania. The office's primary mission is to ensure reliable fossil energy resources for clean, secure, and affordable energy while enhancing environmental protection.
- The Office of Energy Efficiency and Renewable Energy oversees the National Renewable Energy Laboratory in Colorado. The office's mission is to develop solutions for energy-saving homes, buildings, and manufacturing; sustainable transportation; and renewable electricity generation.

Federal agencies other than DOE also provide funding for energy-related R&D. For example, as we found in February 2012, the Department of Defense and USDA implemented numerous initiatives to help develop renewable energy technologies.[28] In addition, as we found in August 2012, the National Aeronautics and Space Administration, National Science Foundation, EPA, and National Institute of Standards and Technology implemented a number of energy initiatives related to batteries and energy storage.[29]

MAJOR FACTORS INFLUENCING U.S. PRODUCTION AND CONSUMPTION OF ENERGY FROM 2000 THROUGH 2013

The following three sections provide information on U.S. production and consumption of fossil, nuclear, and renewable energy from 2000 through 2013 and major factors, including federal activities, that influenced energy production and consumption levels. The fourth section provides information on other federal activities that may have influenced aspects of U.S. energy production and consumption from 2000 through 2013 but were not targeted at a specific energy source, as well as information on federal support for R&D.

Major Factors Influencing U.S. Production and Consumption of Fossil Energy

According to the studies and reports we reviewed, several major factors influenced U.S. production and consumption of fossil energy from 2000 through 2013:

- Advances in drilling technologies enabled economic production of natural gas from shale and other tight formations.[30] These advances led to increases in domestic production of natural gas starting around 2008 and contributed to declines in domestic prices of natural gas starting around 2009. As domestic production rose and prices declined, domestic consumption increased, imports of natural gas decreased, and companies began taking steps to gain approval to export liquefied natural gas.[31]
- The same advances in drilling technologies also enabled the economic production of crude oil from shale formations. These advances led to increases in the domestic production of crude oil beginning around 2009, reversing a decades-long trend of decreasing production. Global crude oil prices generally increased between 2000 and 2013, the largest, sustained price increase since comparable data were available. Increased domestic production contributed to lower prices for some regions of the country; however, the impact of increased domestic crude oil production on global crude oil prices was likely small. Imports of crude oil decreased beginning around 2008 as domestic production displaced imported crude oil to U.S. petroleum refiners. Around 2010, U.S. refiners began consuming greater quantities of crude oil to produce more petroleum products. As domestic consumption of petroleum products generally decreased beginning around 2008, exports of petroleum products (mostly diesel fuel) increased.
- Due in part to lower prices of natural gas, the use of coal for electricity generation decreased in recent years as utilities switched to

natural gas. Domestic coal production decreased in recent years; however, coal exports increased as domestic consumption declined faster than domestic production.

The studies and reports we reviewed also indicated that federal activities may have influenced fossil energy markets, generally by providing incentives or disincentives for the production and consumption of fossil energy. These activities included setting standards and requirements related to fossil energy emissions, assuming risks associated with oil spills and the Strategic Petroleum Reserve, collecting revenues associated with excise taxes on transportation fuels and royalty payments for oil and gas leases, and forgoing revenues associated with tax expenditures for fossil energy producers and royalty relief for oil and gas production. (See app. III for more information on the major factors that influenced U.S. production and consumption of fossil energy from 2000 through 2013.)

Major Factors Influencing U.S. Production and Consumption of Nuclear Energy

According to the studies and reports we reviewed, several major factors may have influenced U.S. production and consumption of nuclear energy from 2000 through 2013. Specifically, declining natural gas prices, along with the 2011 accident at Japan's Fukushima Daiichi commercial nuclear power plant, may have led to decreases in the production and consumption of nuclear energy in recent years.

Federal activities also may have influenced this trend, generally by providing incentives or disincentives for the production and consumption of nuclear energy. These activities included setting standards and requirements related to the operation of nuclear power plants, providing services related to the storage of nuclear waste, assuming risks associated with nuclear power plant operations, and forgoing revenues associated with tax expenditures for nuclear energy producers. (See app. IV for more information on the major factors that influenced U.S. production and consumption of nuclear energy from 2000 through 2013.)

Major Factors Influencing U.S. Production and Consumption of Renewable Energy

According to the studies and reports we reviewed, several major factors influenced U.S. production and consumption of renewable energy—particularly from ethanol, wind energy, and solar energy—from 2000 through 2013:

- Federal tax credits for ethanol and federal policies requiring the use of ethanol in transportation fuels were major factors influencing an 8-fold increase in the production and consumption of ethanol from 2000 through 2013. As domestic production of ethanol outpaced consumption in recent years, U.S. exports of ethanol increased.
- State policies requiring the use of renewable energy in electricity production, as well as federal activities such as outlays and tax credits for renewable energy producers, were major factors influencing production and consumption of electricity from wind and solar energy. Technological advances also played an important role. These factors supported a 30-fold increase in production and consumption of wind energy from 2000 through 2013 and a 19-fold increase in the production and consumption of solar energy.

See appendix V and appendix VI for more information on the major factors that influenced U.S. production and consumption of renewable energy from 2000 through 2013.

Other Federal Activities Influencing Aspects of U.S. Energy Production and Consumption

According to the studies and reports we reviewed, other federal activities were not targeted specifically at fossil, nuclear, or renewable energy production and consumption but may have influenced aspects of U.S. energy production and consumption from 2000 through 2013. Relevant federal activities included setting standards and requirements for energy efficiency, selling electricity, providing loans and loan guarantees related to energy efficiency, making outlays for energy consumption and energy efficiency, and forgoing revenues through tax expenditures for electricity transmission and energy efficiency, among other things. Many of these federal efforts, particularly activities related to energy efficiency, provided disincentives for energy production and use; in contrast, other federal efforts, such as selling electricity, provided incentives for energy production and consumption. In addition, the federal government, and DOE in particular, supported energy-related R&D, which typically is directed at the early stages of technological advances and therefore not generally linked to actual production or consumption of energy. Some of this R&D related to specific energy sources, while other R&D was more general. (See app. VII for more information on federal activities that may have influenced other aspects of U.S. energy production and consumption from 2000 through 2013.)

AGENCY COMMENTS

We provided a draft of this report to DOE, Interior, Treasury, and USDA for review and comment. DOE, Treasury, and USDA provided technical or clarifying comments, which we incorporated as appropriate. Interior indicated they had no comments on the report.

Sincerely yours,
Frank Rusco
Director,
Natural Resources and Environment

APPENDIX I: SCOPE AND METHODOLOGY

This report provides information on U.S. production and consumption of fossil, nuclear, and renewable energy from 2000 through 2013 and major factors, including federal activities, that influenced energy production and consumption levels. It also provides information on other federal activities that may have influenced aspects of U.S. energy production and consumption from 2000 through 2013 but were not targeted at a specific energy source, as well as information on federal support for research and development (R&D).

To provide information on U.S. production and consumption of fossil, nuclear, and renewable energy from 2000 through 2013, we reviewed and analyzed Department of Energy (DOE) Energy Information Administration (EIA) historical data, as well as EIA articles and monthly and annual reports.

To assess the reliability of EIA data, we took several steps including reviewing available documentation on the collection of the data. We determined the EIA data to be sufficiently reliable for our purposes.

To provide information on the major factors, including federal activities, that influenced energy production and consumption levels from 2000 through 2013; to provide information on other federal activities that may have influenced U.S. energy production and consumption from 2000 through 2013 but were not targeted at a specific energy source; and to provide information on federal R&D, we performed the following steps:

- We reviewed information from our prior work related to the production and consumption of crude oil, petroleum products, natural gas, coal, nuclear energy, ethanol, wind energy, solar energy, energy efficiency, electricity transmission, and federal R&D. We relied on our prior work to identify the major factors that influenced U.S. energy production and consumption levels. However, this work may not have identified all of the relevant factors that influenced energy production and consumption from 2000 through 2013. In addition, we did not examine permitting issues on federal lands, including issues related to infrastructure and oil and gas development.
- We reviewed information related to U.S. energy production and consumption from federal agencies and government organizations, including reports and studies by the Congressional Budget Office (CBO), Congressional Research Service (CRS), EIA, congressional Joint Committee on Taxation (JCT), and U.S. Department of Agriculture (USDA). To identify reports and studies, we conducted searches of various databases, such as ProQuest and PolicyFile, for studies published since 2008. We also asked agency officials and other stakeholders we contacted to recommend reports and studies. We relied on these reports and studies to identify the major factors that influenced U.S. energy production and consumption levels. However, these reports and studies may not have identified all of the relevant factors that influenced energy production and consumption from 2000 through 2013.
- We reviewed and analyzed data on outlays, royalties collected, and excise taxes collected from DOE, Department of the Interior (Interior), Department of the Treasury (Treasury), and Office of Management and Budget (OMB). To assess the reliability of these data sets, we interviewed individuals with knowledge of them and

reviewed available documentation on the collection of the data and on any methods that were used in calculating the data. From this review, we determined that the data sets were sufficiently reliable for our purposes.

- We reviewed and analyzed data on estimates of tax expenditures, forgone royalties, and federal credit programs collected from DOE, JCT, and Treasury.[1] We also relied on lists of tax expenditures and estimates of their cost compiled annually by Treasury and JCT under the energy budget function. In general, we used Treasury revenue loss estimates for each tax expenditure except in cases where only JCT reported a tax expenditure.[2] Regarding data on tax expenditures, changes in economic conditions and estimation techniques can affect revenue loss estimates for tax expenditures, making them differ from year to year. Also, legislation affecting tax rates or the tax structure affects tax expenditure estimates. When statutory rates increase, a taxpayer's ability to reduce tax on a portion of income is worth more; consequently, tax expenditures are worth more. Likewise, when rates decrease, tax expenditures are worth relatively less. To assess the reliability of these data sets, we interviewed individuals with knowledge of them and reviewed available documentation on the collection of the data and on any methods that were used in calculating the data. From this review, we determined that the data sets were sufficiently reliable for our purposes.

We conducted this performance audit from August 2013 to September 2014 in accordance with generally accepted government auditing standards. Those standards require that we plan and perform the audit to obtain sufficient, appropriate evidence to provide a reasonable basis for our findings and conclusions based on our audit objectives. We believe that the evidence obtained provides a reasonable basis for our findings and conclusions based on our audit objectives.

APPENDIX II: FEDERAL SOURCES OF DATA ON ENERGY-RELATED OUTLAYS AND TAX EXPENDITURES

The federal budget provides data on energy-related outlays. The federal budget is divided into different functional areas, which is a way of grouping budgetary resources so that all budget authority and outlays of on-budget and off-budget federal entities and tax expenditures can be presented according to the national needs being addressed. National needs are grouped in 17 broad areas, one of which is “energy.” The energy area includes (1) promoting an adequate supply and appropriate use of energy to serve the needs of the economy and (2) energy programs of the Department of Energy (DOE) and its predecessor agencies. It also excludes atomic energy defense activities and general science research not closely related to energy. The Office of Management and Budget’s (OMB) public budget database contains historical data on outlays associated with each functional area, including the energy area.

However, the federal budget does not provide a comprehensive source for all federal outlays that might be related to energy production and consumption. Instead, the 17 functional areas in the federal budget relate to the primary area of a given account, even though the programs within the account may serve a variety of purposes.

As a result, the public budget database may not identify all energy-related activities in the energy functional area. For example, the federal budget identifies seven agencies, including DOE and the U.S. Department of Agriculture (USDA), with outlays in the energy supply subfunction in fiscal year 2010. However, as we found in February 2012, 17 federal agencies beside DOE and USDA—such as the Departments of Defense and the Interior—implemented renewable energy initiatives for fiscal year 2010.[1]

The Department of the Treasury (Treasury) and the congressional Joint Committee on Taxation (JCT) report revenue loss estimates for energy-related tax expenditures.

Both Treasury and JCT estimate the revenue loss[2] associated with each tax provision they have identified as a tax expenditure.[3] Treasury’s list is included in the President’s annual budget submission; and JCT issues annual tax expenditure estimates as a standalone product. Both organizations calculate a tax expenditure as the difference between tax liability under current law and what the tax liability would be if the provision were eliminated and the item were treated as it would be under a “normal” income tax. In general, the tax expenditure lists that Treasury and JCT publish are similar, although these lists differ somewhat in the number of tax expenditures reported and the estimated revenue losses for particular expenditures. In addition, as with the federal budget, both lists of tax expenditures are divided into different functional areas, including one related to energy.

However, Treasury’s list and JCT’s list of tax expenditures may not include all tax expenditures that provide a benefit to energy producers. For example, both Treasury and JCT list the deduction for income attributable to domestic production activities as a tax expenditure. This tax expenditure allows a deduction of 6 percent from taxable income for oil extraction, among other things.

According to Treasury estimates, repealing this provision would result in more than $17 billion in revenue related to oil and natural gas production for fiscal years 2014 through 2023. However, because this tax expenditure is available to other industries beside oil production, Treasury and JCT do not list this tax expenditure under the energy functional area.

Appendix III: Information on Major Factors Influencing Fossil Energy Production and Consumption

This appendix provides more detailed information on U.S. production and consumption of fossil energy from 2000 through 2013 and major factors, including federal activities, that influenced fossil energy production and consumption levels.

Natural Gas Production and Consumption

The year-to-year pattern of domestic production of natural gas fluctuated from 2000 through 2006 and then began to increase around 2007, according to Energy Information Administration (EIA) data, and as shown in figure 3. Specifically, the United States produced about 19.2 trillion cubic feet of natural gas in 2000; by 2005 and 2006, production had fallen below 19 trillion cubic feet but then began to increase, reaching over 24 trillion cubic feet in 2012 and 2013. Domestic consumption of natural gas exceeded domestic production throughout the period, with the difference coming from imports, primarily from Canada.[1]

However, as shown in figure 3, the difference between the domestic consumption and production of natural gas generally decreased between 2007 and 2013, leading to a reduction in natural gas imports.

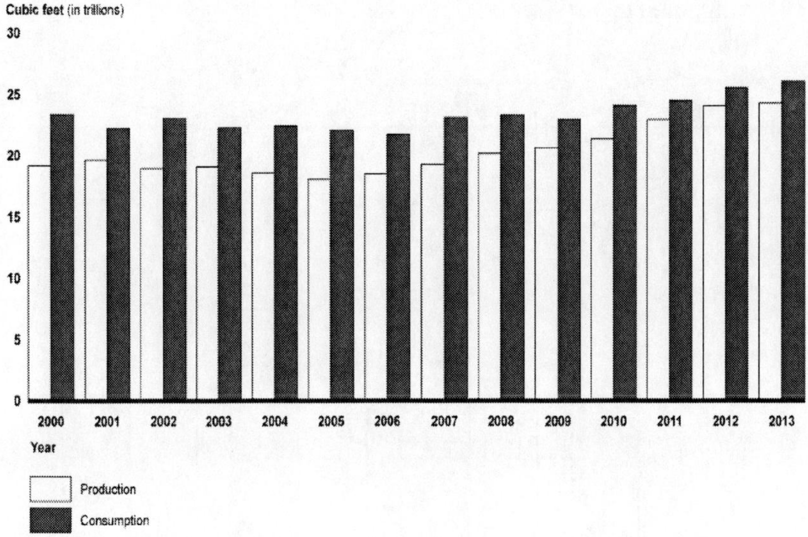

Source: GAO analysis of Energy Information Administration data. | GAO-14-836.

Note: Production refers to "dry" production of natural gas, which is the natural gas that remains after the liquid constituents (i.e., natural gas plant liquids) have been removed.

Figure 3. U.S. Natural Gas Production and Consumption, 2000-2013.

Natural gas is used by a number of sectors in the economy, most notably for electricity generation; for industrial use as a source of heat or as a feedstock for petrochemical production, among other things; for residential heating and other home uses; and for commercial heating and other uses. Figure 4 shows the share of natural gas consumption by sector in 2000 and 2013.

Specifically, according to EIA data, natural gas consumption for electricity generation (as well as other energy needs of the electric power sector) increased from about 5.2 trillion cubic feet in 2000 to about 8.2 trillion cubic feet in 2013. Natural gas consumption for commercial use also increased, from about 3.2 trillion cubic feet in 2000 to about 3.3 trillion cubic feet in 2013. Industrial and residential uses declined over the same period.

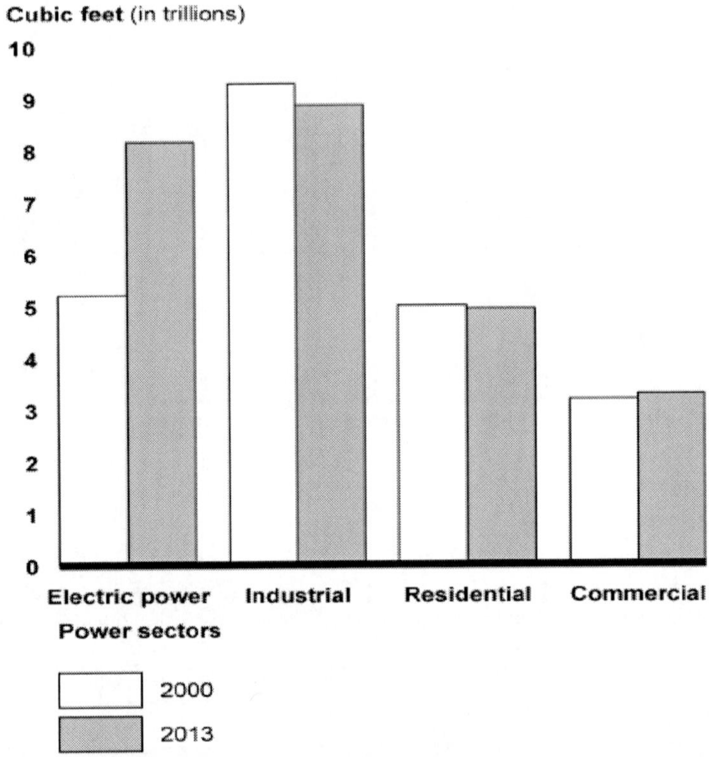

Source: GAO analysis of Energy Information Administration data. | GAO-14-836.

Note: The transportation sector consumes about 3 percent of annual natural gas production and is excluded from this chart.

Figure 4. U.S. Natural Gas Consumption by Sector, 2000 and 2013.

Major Factors Influencing Natural Gas Production and Consumption

The studies and reports we reviewed indicated that increases in the domestic production of natural gas were due primarily to increases in the extraction of natural gas from shale formations. As shown in figure 5, the production of natural gas from shale formations caused an increase in total domestic natural gas production starting around 2008 and continuing through 2012 (the latest year for which annual data were available). According to EIA data, natural gas withdrawals from shale formations increased from about 2 trillion cubic feet in 2007 to over 10 trillion cubic feet in 2012. These increases were largely due to technological advances in horizontal drilling and hydraulic fracturing—a process that injects a combination of water, sand, and chemical additives under high pressure to create and maintain fractures in underground rock formations that allow oil and natural gas to flow. For example, as we reported in January 2012, improvements in horizontal drilling and hydraulic fracturing led to a boom in the production of natural gas from shale formations.[2] In addition, according to the Congressional Research Service (CRS), in recent years the oil and gas industry improved its extraction rate of natural gas from shale formations from about 5 percent to about 15 percent of total estimated gas resources in the ground (thereby tripling the amount of recoverable natural gas).[3]

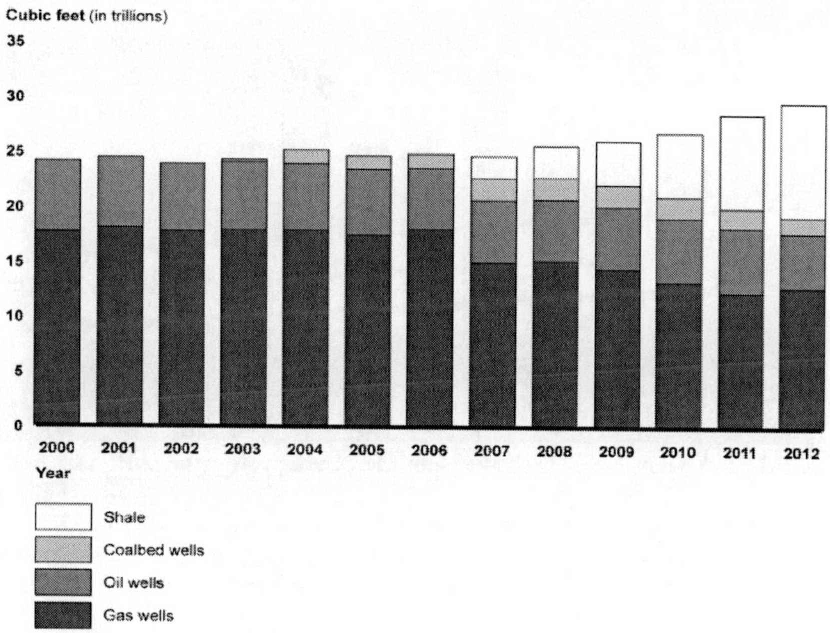

Source: GAO analysis of Energy Information Administration data. | GAO-14-836.

Note: Sources of natural gas production include gross withdrawals of natural gas from oil wells, gas wells, coalbed wells, and shale formations. Gross withdrawals include liquid constituents (i.e., natural gas plant liquids) and dry constituents of natural gas. 2012 is the latest year for which data were available.

Figure 5. Sources of U.S. Natural Gas Production, 2000-2012.

Increases in domestic natural gas production contributed to decreases in domestic natural gas prices, according to CRS.[4] The price of natural gas, as with other commodities, is driven by supply and demand. The Henry Hub spot market in Louisiana is the best known spot market for natural gas.[5] As shown in figure 6, annual prices for natural gas in the Henry Hub spot market generally increased between 2000 and 2008 (although some fluctuations occurred) before decreasing between 2008 and 2013. Specifically, in 2000, the annual spot price was $4.31 per million British thermal units (Btu) of natural gas. This price generally increased to $8.69 per million Btus in 2005 and $8.86 per million Btus 2008. Since 2008, the annual price generally decreased to $3.73 per million Btus in 2013.

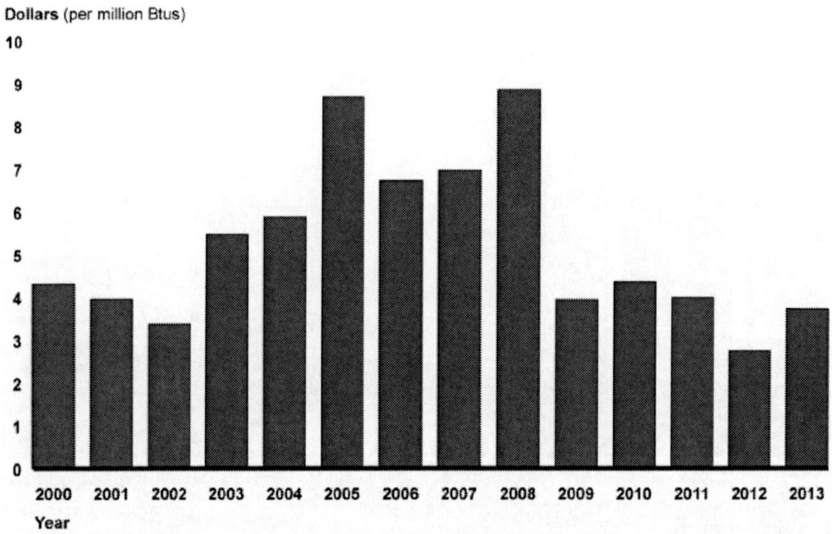

Source: GAO analysis of Energy Information Administration data. | GAO-14-836.

Note: Btu stands for British thermal units. Prices refer to average annual Henry Hub natural gas spot prices. Data are not adjusted for inflation.

Figure 6. U.S. Natural Gas Prices, 2000-2013.

In recent years, natural gas prices in the United States were much lower than in other parts of the world, according to CRS.[6] This price difference encouraged some American companies to apply for authorization to export domestically produced liquefied natural gas from the contiguous 48 states. Specifically, since 2010, the Department of Energy (DOE) has received more than 30 applications for permission to export liquefied natural gas to countries that do not have a free trade agreement with the United States.[7] DOE has fully approved 3 applications and approved 6 others on the condition that the Federal Energy Regulatory Commission issues a satisfactory environmental review of the associated liquefied natural gas export facility.[8] Moreover, EIA has projected that the United States could be a net exporter of liquefied natural gas by 2016.[9]

Production and Consumption of Crude Oil and Petroleum Products

The United States produces crude oil, which is refined along with imported crude oil into petroleum products such as gasoline, diesel, and jet fuel. U.S. refineries use both domestically produced crude oil and imported crude oil to produce petroleum products. According to EIA data, there were 143 petroleum refineries in the United States as of January 2013, with a capacity to process 17.8 million barrels of crude oil per day.[10] The United States also both exports and imports petroleum products.

Domestic production of crude oil declined from 2000 through 2008, continuing a downward trend beginning in the 1970s, but increased beginning in 2009, according to EIA data. Specifically, as shown in figure 7, between 2000 and 2008, domestic crude oil production decreased from an average of about 5.8 million barrels of oil per day to 5.0 million barrels per day. However, by 2013, domestic crude oil production had increased to an average of almost 7.5 million barrels per day, the highest level of oil production since 1989. In 2013, an average of about 15.3 million barrels per day of crude oil was refined into petroleum products at U.S. refineries. As domestic crude oil production increased, imports of foreign crude oil decreased. Specifically, crude oil imports decreased from an average of about 9.8 million barrels per day in 2008 to 7.7 million barrels per day in 2013.

Overall domestic consumption of petroleum products—including gasoline, diesel, and jet fuel—peaked in 2005 and then generally declined through 2013, according to EIA data. Specifically, as shown in figure 8, consumption of petroleum products increased from an average of 19.7 million barrels per day in 2000 to 20.8 million barrels per day in 2005 before declining to between 18-19 million barrels per day in recent years. Amid declining domestic consumption and somewhat increasing U.S. refining capacity, refiners have increasingly exported petroleum products. Specifically, U.S. exports of petroleum products grew from an average of about 1.1 million barrels per day in 2005 to about 3.5 million barrels per day in 2013.[11] The United States had been a net importer of petroleum products since 1949, but in 2011 it became a net exporter, primarily of diesel.

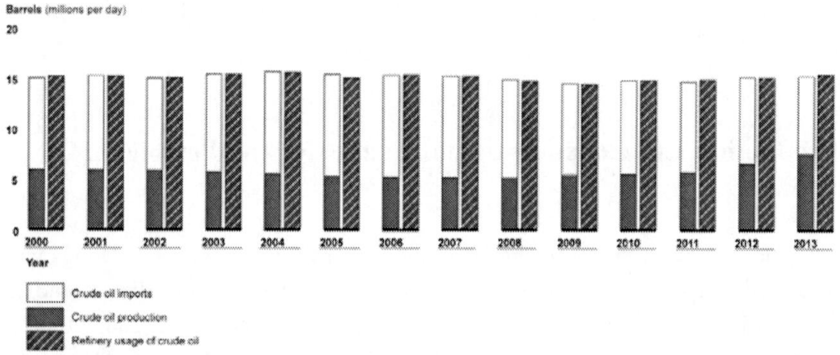

Source: GAO analysis of Energy Information Administration data. | GAO-14-836.

Note: Refinery usage of crude oil includes domestically produced crude oil and imported crude oil, as well as usage from domestic inventories of crude oil.

Figure 7. U.S. Crude Oil Production and Usage of Crude Oil at U.S. Refineries, 2000-2013.

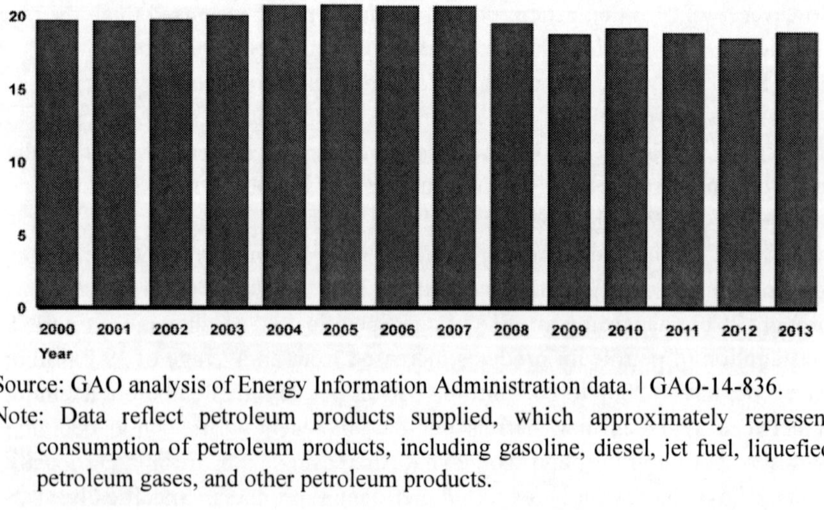

Source: GAO analysis of Energy Information Administration data. | GAO-14-836.

Note: Data reflect petroleum products supplied, which approximately represent consumption of petroleum products, including gasoline, diesel, jet fuel, liquefied petroleum gases, and other petroleum products.

Figure 8. U.S. Consumption of Petroleum Products, 2000-2013.

The primary users of petroleum products in the United States are the transportation and industrial sectors, according to EIA data. As shown in figure 9, the transportation sector consumed the largest share of petroleum products in 2000 and 2013 (at about 4.8 billion barrels). The industrial sector consumed the next largest share of petroleum products (at about 1.8 billion barrels in 2000 and about 1.7 billion barrels in 2013), while the remaining sectors (commercial, residential, and electric power) consumed the smallest share (at about 0.7 billion barrels in 2000 and about 0.3 billion barrels in 2013). Overall, total consumption of petroleum products increased from about 7.2 billion barrels in 2000 to about 7.6 billion barrels in 2007 before decreasing to about 6.9 billion barrels in 2013.

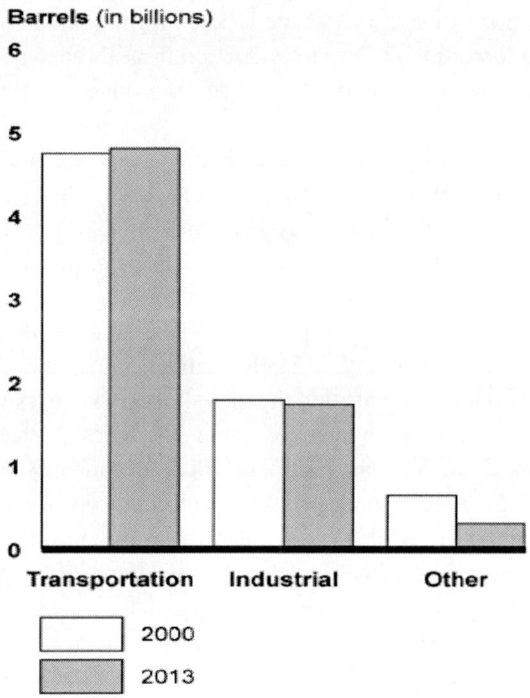

Source: GAO analysis of Energy Information Administration data. | GAO-14-836.
Note: Other includes the commercial, residential, and electric power sectors.

Figure 9. U.S. Consumption of Petroleum Products by Sector, 2000 and 2013.

Major Factors Influencing Production and Consumption of Crude Oil and Petroleum Products

The studies and reports we reviewed indicated that increases in domestic crude oil production came primarily from increased production from shale formations. As we found in March 2014, the technological advances in horizontal drilling and hydraulic fracturing that contributed to increasing U.S. production of natural gas have also allowed companies that develop petroleum resources to extract crude oil from shale and other formations that were previously considered to be inaccessible because traditional techniques did not yield sufficient amounts for economically viable production.[12] According to EIA data, much of the increase in crude oil production has been from shale and other tight formations, such as the Bakken formation in North Dakota and the Eagle Ford formation in Texas. Production from these two states accounted for 87 percent of the increase in U.S. crude oil production from 2008 through 2013.

Prices of crude oil increased considerably between 2000 and 2013 but show a distinctive pattern, according to EIA data. First, crude oil prices generally increased from 2000 to 2008. Crude oil is a widely traded global commodity, and prices of crude oil are generally determined by global supply and demand rather than exclusively by events in a single oil-producing country such as the United States. Since the mid-1980s, benchmark crude oil prices such as West Texas Intermediate in the United States and Brent in Europe have served as reference points that the global oil market uses for pricing other crude oils.[13] As shown in figure 10, crude oil prices in these two markets increased from about $30 per barrel in 2000 to almost $100 per barrel in 2008—one of the largest and most sustained crude oil price increases since comparable data were available. In 2009, crude oil prices in both markets decreased to about $62 per barrel, reflecting the effects of the global economic recession of 2007 to 2009 and the consequent falling demand for petroleum products. In 2010, crude oil prices in both markets increased.

Starting in 2010, increases in U.S. production of crude oil contributed to lower prices for some domestic crude oils and may have had a small effect on some global oil prices, according to EIA.[14] Specifically, as shown in figure 10, West Texas Intermediate crude oil began to sell at a large discount relative to Brent crude oil starting around 2011. This price divergence was due to increases in domestic crude oil production— existing crude oil pipelines were constrained in their ability to ship additional quantities of oil from the midcontinental United States and Canada to refineries, which limited the ability of this oil to reach global oil markets.[15] In terms of global effects, in 2013 U.S. crude oil production grew more than the combined increase in the rest of the world, which contributed to relatively stable global crude oil prices in 2013, according to EIA.[16]

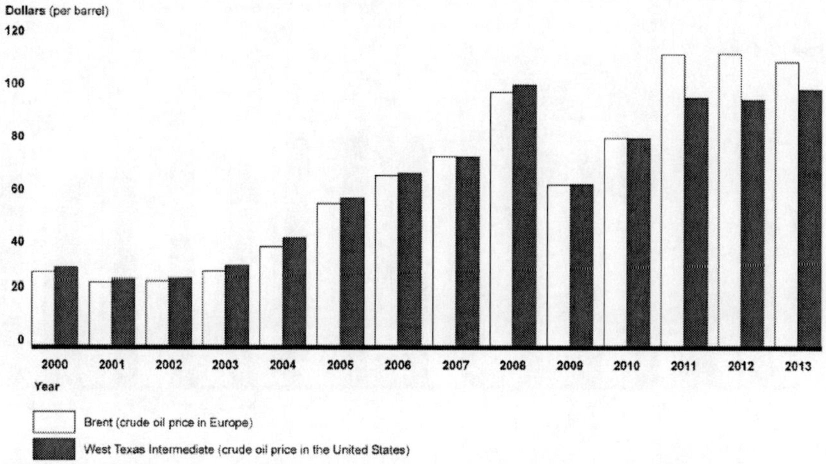

Source: GAO analysis of Energy Information Administration data. | GAO-14-836.

Note: Prices refer to average annual spot prices. A spot price is a price for one-time open market transactions for immediate delivery of a specific quantity of product at a specific location where the commodity is purchased at current market rates. Data are not adjusted for inflation.

Figure 10. Crude Oil Prices, 2000-2013.

Coal Production and Consumption

The United States has the largest recoverable coal reserves in the world, according to EIA, and most domestically produced coal comes from five states: Wyoming, West Virginia, Kentucky, Pennsylvania, and Illinois.[17] Domestic production of coal fluctuated somewhat during the past decade but declined by about 16 percent from 2008 through 2013, according to EIA data. Specifically, as shown in figure 11, the United States produced 1.07 billion short tons of coal in 2000;[18] by 2008, production had risen to 1.17 billion short tons. However, coal production fell below 1 billion short tons by 2013—the lowest level in almost 2 decades, according to EIA data. Domestic consumption of coal generally increased from 2000 (at 1.08 billion short tons) to 2007 (at 1.13 billion short tons). However, starting in 2008, coal consumption generally decreased and reached 0.93 billion short tons in 2013. As domestic consumption of coal fell, the United States generally exported more coal to Europe and Asia. The vast majority of coal consumed in the United States is used for generating electricity.

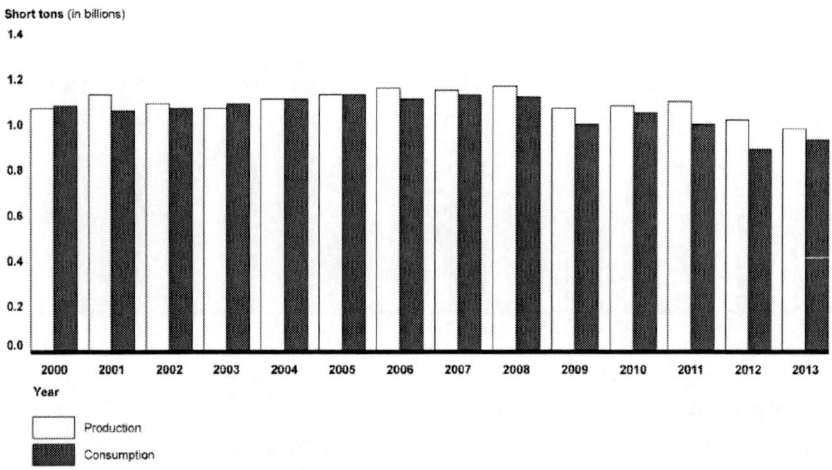

Source: GAO analysis of Energy Information Administration data. | GAO-14-836.
Note: A short ton is 2,000 pounds.

Figure 11. U.S. Coal Production and Consumption, 2000-2013.

Major Factors Influencing Coal Production and Consumption

The studies and reports we reviewed indicated that the recent decrease in domestic coal production and consumption came partly from declines in the use of coal to generate electricity. In the past, the fuel cost of generating one kilowatt-hour[19] of electricity from natural gas had typically been higher than that of coal, according to EIA.[20] However, coal began losing its price advantage over natural gas for electricity generation in some parts of the country in 2009, particularly in the eastern United States, according to EIA. In addition, new natural gas-fueled generating units generally are able to convert fuel into electricity more efficiently than existing coal-fueled generating units, meaning they can convert a unit of fuel energy into more electricity than coal-fueled units.[21] Newer designs of coal-fueled units exist that can operate at higher efficiencies, but few have been built in the United States. In addition, recently proposed or finalized EPA regulations affecting coal-fueled electricity generating units may also have played a role in recent decreases in domestic coal production and consumption.[22]

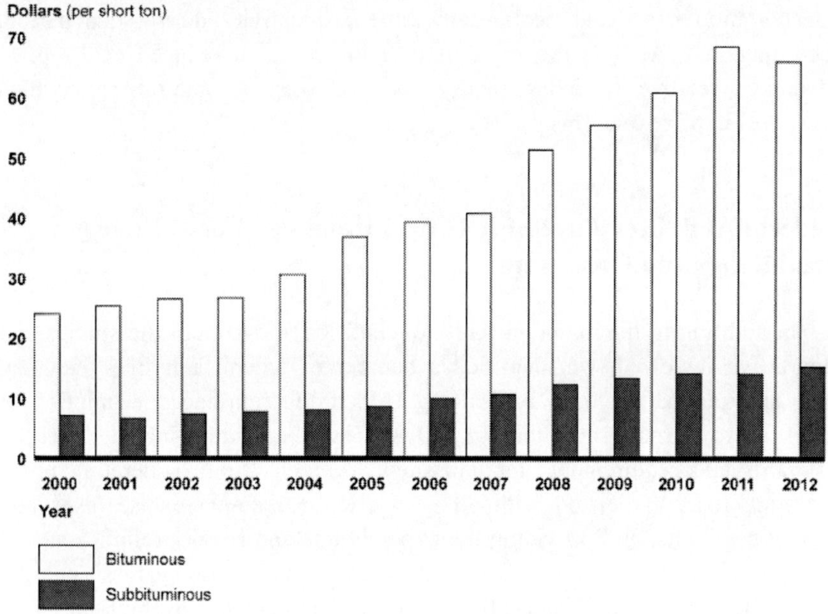

Source: GAO analysis of Energy Information Administration data. | GAO-14-836.

Note: Prices are annual free-on-board prices—that is, prices at the coal mine or prices at the producer country's port of loading—before the cost of insurance, freight, and credit. 2012 is the latest year for which data were available. Data are not adjusted for inflation.

Figure 12. U.S. Coal Prices, 2000-2012.

The price of coal depends on its type, in part because different types of coal produce differing amounts of energy when burned. According to EIA, two of the most common types in the United States are bituminous and subbituminous coal.[23] Bituminous coal is the oldest and most abundant coal type found in the United States. West Virginia, Kentucky, and Pennsylvania are the primary producers of bituminous coal. Subbituminous coal contains less energy than bituminous coal; however, large quantities are found in thick beds near the surface, resulting in low mining cost and, correspondingly, lower prices. Wyoming produces the vast majority of subbituminous coal. As shown in figure 12, annual U.S. prices for bituminous and subbituminous coal generally increased from 2000 to 2012 (the latest year for which data are available). For example, for bituminous coal, prices increased from $24.15 per short ton in 2000 to $66.04 per short ton in 2012, or an increase of over 170 percent. Some of these cost increases may be due to increases in coal transportation costs and declines in mine productivity during this period, according to EIA.[24] As the price of coal increased, it reduced coal's price advantage relative to other energy sources, such as natural gas, which decreased in price over this period.

Federal Activities That May Have Influenced Fossil Energy Production and Consumption

In addition to the major factors we identified above from the studies and reports we reviewed, we identified a number of federal activities that may have also played a role in influencing U.S. production and consumption of fossil energy from 2000 through 2013. These activities included setting standards and requirements for emissions from electricity generating units, assuming risks associated with oil production, collecting excise taxes and royalty payments, and providing tax expenditures and royalty relief. Some of these activities provided an incentive to produce and consume fossil energy, while others provided a disincentive for its production and consumption.

Setting Standards and Requirements

The federal government established or strengthened a number of standards and requirements related to fossil energy production from 2000 through 2013. For example, under the Clean Air Act,[25] the Environmental Protection Agency (EPA) establishes national ambient air quality standards for six pollutants which states are, primarily, responsible for attaining.[26] States attain these standards, in part, by regulating emissions of these pollutants from certain stationary sources, such as electricity generating units. In particular, according to EPA, fossil fuel-fired electricity generating units are among the largest emitters of sulfur dioxide and nitrogen oxides, which have been linked to respiratory illnesses and acid rain, as well as of carbon dioxide, the primary greenhouse gas contributing to climate change. Numerous Clean Air Act requirements apply to electricity generating units, including New Source Review, a permitting process established in 1977. Under New Source Review, owners of generating units must obtain a preconstruction permit that establishes emission limits and requires the use of certain emissions control technologies. New Source Review applies to (1) generating units built after August 7, 1977, and (2) existing generating units—regardless of the date built—that seek to undertake a "major modification," a physical or operational change that would result in a significant net increase in emissions of a regulated pollutant.

In general, the cost of complying with New Source Review requirements provided a disincentive for producing electricity from fossil energy sources. As we found in June 2012, EPA has investigated most coal-fired generating units at least once for compliance with New Source Review requirements since 1999, and has alleged noncompliance at more than half of the units it investigated.[27] Specifically, of the 831 units EPA investigated, 467 units were ultimately issued notices of violation, had complaints filed in court, or were included in settlement agreements. In total, EPA reached 22 settlements covering 263 units, which will require affected unit owners to, among other things, install around $12.8 billion in emissions controls. According to our analysis of EPA data, these settlements will reduce emissions of sulfur dioxide by an estimated 1.8 million tons annually, and nitrogen oxides by an estimated 596,000 tons annually.

Assuming Risk

The federal government assumed some risks related to fossil energy production and consumption from 2000 through 2013. For example, the federal government assumed financial risks associated with potential cleanup costs for some oil spills, and the federal government acquired billions of dollars worth of crude oil to hold in reserve in case of supply disruptions, as discussed below:

- **Cleanup costs for oil spills.** Under the Oil Pollution Act of 1990, as amended, which was enacted after the Exxon Valdez oil spill in 1989, the federal government established a "polluter pays" system that places the primary burden of liability for costs of spills on the responsible parties, up to a specified limit of liability.[28] In general, the level of potential financial liability under the act depends on the kind of vessel or facility from which a spill originates and is limited in amount. However, if the oil discharge is the result of gross negligence or willful misconduct, or a violation of federal operation, safety, and construction regulations, then liability under the act is unlimited. In addition, the act provides the Oil Spill Liability Trust Fund to pay for oil spill costs when the responsible party cannot or does not pay.[29]

The fund's primary revenue source is an 8-cent-per-barrel tax on petroleum products—a small fraction of the price of a barrel in 2013— either produced in the United States or imported from other countries.[30] The fund is subject to a $1 billion cap on the amount of expenditures from the fund per incident.

- **Stockpiling crude oil.** Congress created the Strategic Petroleum Reserve in 1975, following the Arab oil embargo of 1973 to 1974, to help protect the U.S. economy from damage caused by oil supply disruptions. The reserve is owned by the federal government and operated by DOE. It can store up to 727 million barrels of crude oil in salt caverns. The President has discretion to authorize release of oil in the Strategic Petroleum Reserve to minimize significant supply disruptions.[31] In the event of such a disruption, the reserve can supply oil to the market by either selling stored crude oil or trading this oil in exchange for a larger amount of oil to be returned later. From fiscal year 2000 through 2013, the federal government received almost $3.9 billion from the sale of crude oil from the reserve, spent about $0.5 billion to purchase crude oil, and spent $2.5 billion for operations and maintenance of the reserve.[32]

The assumption of liability by the federal government for some oil spills may have provided an incentive for oil production and consumption by potentially decreasing the overall cost associated with certain production-related activities. For example, the liability limitations established under the Oil Pollution Act may have lowered costs for liability insurance or other insurance paid for by oil producers. However, the extent to which this federal intervention influenced changes in petroleum or natural gas production or consumption is difficult to precisely measure. Moreover, the fund—which is paid by oil producers—raises the cost of producing oil by a small fraction, which may have a negative impact on oil production.[33]

DOE's operation of the Strategic Petroleum Reserve may have had little to no impact on fossil energy production and consumption between 2000 and 2013. For example, we found in August 2006 that filling the Strategic Petroleum Reserve from late 2001 through 2005 during a time of tight supply and demand conditions had minimal impact on oil prices because the volume was so small compared with world oil demand, according to most of the experts with whom we spoke.[34] However, some experts believed the existence of the Strategic Petroleum Reserve may have a stabilizing effect on oil prices, particularly during extreme supply or demand events, whether or not it is actually used. If true, the reserve could have had positive effects on oil producers or consumers by reducing the risks associated with unstable prices.

Collecting or Forgoing Revenue

From 2000 through 2013, the federal government collected revenues through excise taxes and royalty payments related to fossil energy production and consumption while forgoing other related revenues through tax expenditures and royalty relief. Regarding excise taxes, the federal government collected about $637 billion through excise taxes targeting or related to fossil energy—primarily motor fuels (gasoline, diesel, and others)—from fiscal year 2000 through 2012.[35] The federal excise tax rate on gasoline is 18.4 cents per gallon (the same amount as in 1993). Most revenues from these taxes are dedicated to the Highway Trust Fund, which was established by Congress in 1956 and is a major source of funding for various surface transportation programs.[36] As shown in figure 13, revenues from excise taxes targeting or related to fossil energy were about $45 billion a year from fiscal year 2000 through 2004 and increased to about $50 billion a year for the rest of the period.

Regarding royalty payments, the federal government collected more than $124 billion in revenues from royalty and other payments for federal oil, gas, and coal leases from fiscal year 2003 through 2013.[37] As shown in figure 14, revenues from royalty and other payments increased from almost $8 billion in fiscal year 2003 to $23.4 billion in fiscal year 2008, then decreased to about $9 billion in fiscal years 2009 and 2010 before increasing to about $13.2 billion in fiscal year 2013.

Regarding tax expenditures, the federal government incurred revenue losses of almost $50 billion from fiscal year 2000 through 2013 due to 16 tax expenditures we identified as targeting or related to fossil energy according to the Department of the Treasury (Treasury) and the Joint Committee on Taxation (JCT) estimates. As shown in figure 15, revenue losses associated with these 16 tax expenditures increased from less than $2 billion in fiscal year 2000 to over $4.6 billion in both fiscal year 2006 and 2007. They decreased from fiscal year 2007 through 2010 before increasing to about $4.6 billion in fiscal year 2012 and declining to $4.1 billion in fiscal year 2013.

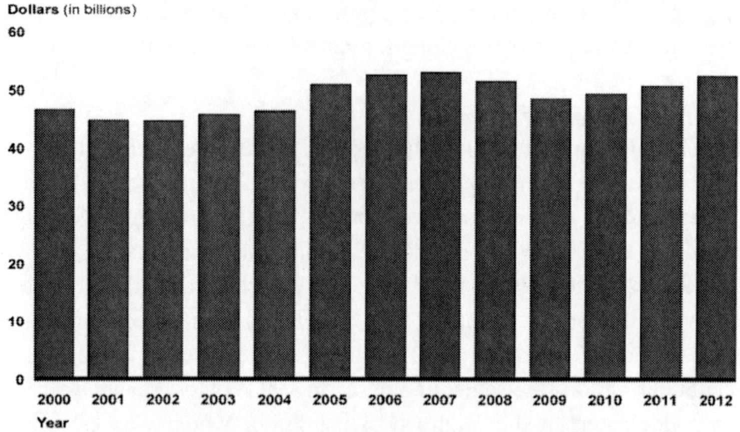

Source: GAO analysis of Internal Revenue Service data. | GAO-14-836.

Note: The most recent data available from the Internal Revenue Service on excise taxes collected are for 2012. These data include excise tax collections from sales of motor fuels (gasoline, diesel, kerosene, aviation-grade kerosene, liquefied natural gas, liquefied fuel derived from coal, liquefied hydrogen, compressed natural gas, liquid petroleum gas, and others), passenger vehicles that are less fuel-efficient (the "gas-guzzler" tax), trucks and tires, coal, passenger transportation by air and water, and others. Data are not adjusted for inflation.

Figure 13. Revenue from Federal Excise Taxes Targeting or Related to Fossil Energy, Fiscal Year 2000 – 2012.

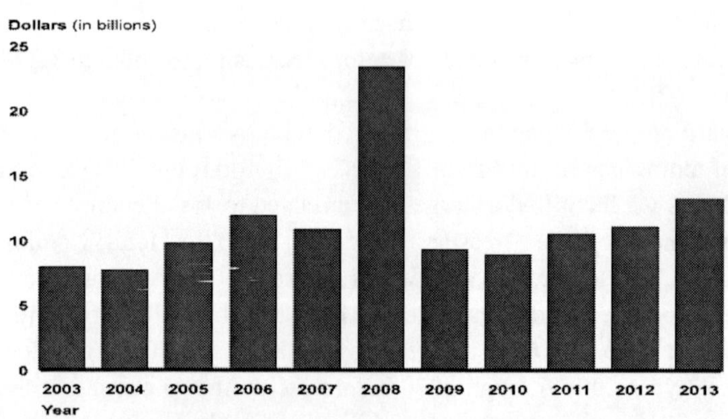

Source: GAO analysis of Interior data. I GAO-14-836.

Note: According to Department of the Interior (Interior) documentation, comparable data were not available for 2000 through 2002, so data on royalty and other payments are not included for those years. Data are not adjusted for inflation.

Figure 14. Revenue from Royalty and Other Payments for Federal Oil, Gas, and Coal Leases, Fiscal Year 2003-2013.

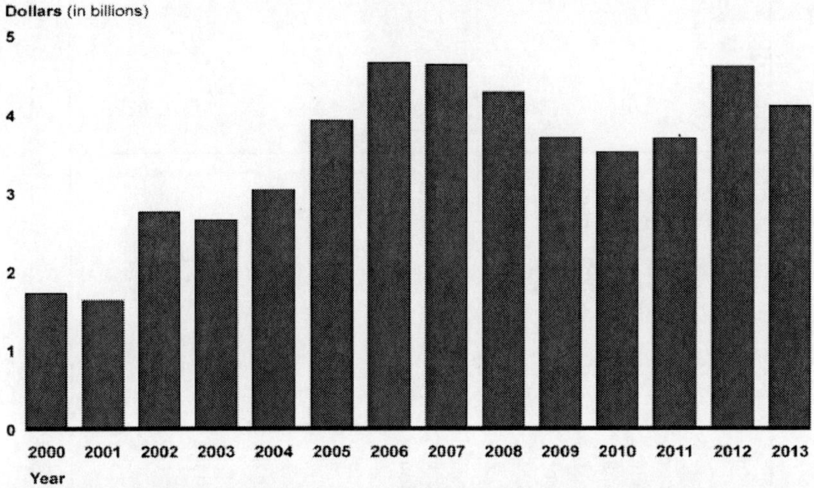

Source: GAO analysis of Treasury and Joint Committee on Taxation data. | GAO-13-836.

Note: Summing tax expenditure estimates is a useful gauge of size but does not take into account possible interactions among individual tax expenditures. Data are not adjusted for inflation.

Figure 15. Estimated Revenue Losses Associated with Federal Tax Expenditures Targeting or Related to Fossil Energy, Fiscal Year 2000 – 2013.

Table 2 provides descriptions of these 16 federal tax expenditures. The table also provides information from Treasury on tax expenditures that will or have expired, in full or in part, due to an expiration of legislative authority or some other expiration under the law as of the fall of 2014, as well as on tax expenditures that currently have no expiration. In addition, the table provides information on revenue loss estimates from Treasury (unless otherwise specified). The federal government may have incurred additional revenue losses associated with other tax provisions related to fossil energy; however, we were unable to identify data on the level of federal revenue losses. For example, JCT and Treasury identified three tax provisions as being related to the oil and gas industry but also broadly available to taxpayers engaged in energy-related and non-energy-related activities, such as manufacturing or trade. These three tax provisions are described in table 3.[38] For these provisions, JCT and Treasury did not estimate annual revenue losses attributable to fossil energy production or consumption for fiscal year 2000 through 2013, and therefore this information was not readily available.

Table 2. 16 Federal Tax Expenditures Targeting or Related to Fossil Energy, Fiscal Year 2000 – 2013

Name	Description	Expirationinformation	Estimated revenue losses
Nonconventional fuel credit (also called alternative fuel production credit)a	The tax code provides a credit of $3 per oil-equivalent barrel of production (in 1979 dollars) for certain types of liquid, gaseous, and solid fuels produced from selected types of alternative energy sources (or "nonconventional fuels"). Qualifying fuels include synthetic fuels (such as coke or coke gas) produced from coal, as well as gas produced from biomass, among other things. The credit is generally available if the price of oil stays below $29.50 (in 1979 dollars).	Qualifying facilities producing coke or coke gas must be placed in service by December 31, 2009.	$14.83 billion
Excess of percentage over cost depletion, fuelsb	The tax code allows firms that extract oil, gas, or other minerals a deduction to recover their capital investment in a mineral reserve, which depreciates due to the physical and economic depletion or exhaustion as the mineral is recovered. There are two methods of calculating this deduction: cost depletion and percentage depletion. Cost depletion allows for the recovery of the actual capital investment—the costs of discovering, purchasing, and developing a mineral reserve—over the period during which the reserve produces income from the specified total recoverable units. Under this method, the total deductions cannot exceed the original capital investment. Under percentage depletion, the deduction for recovery of capital investment is a fixed percentage of the "gross income"—i.e., revenue—from the sale of the mineral.	No expiration under current law.	$10.15 billion

Name	Description	Expirationinformation	Estimated revenue losses
	Because eligible taxpayers must claim the higher of cost or percentage depletion, total deductions under percentage depletion may exceed the capital invested to acquire and develop the reserve. The percentage depletion rate for oil and gas is 15 percent and is limited to average daily production of 1,000 barrels of oil, or its equivalent in gas, and only for wells located in the United States. Percentage depletion is available for independent producers and royalty owners but not for integrated producers.		
Expensing of exploration and development costs, fuelsc	Firms engaged in the exploration and development of oil, gas, or geothermal properties have the option of expensing rather than capitalizing certain intangible drilling and development costs. Intangible drilling and development costs are amounts paid by the operator for fuel, labor, repairs to drilling equipment, materials, hauling, and supplies. They are expenditures incidental to and necessary for drilling wells and preparing a site for the production of oil, gas, or geothermal energy. They include the cost to operators of any drilling or development work done by contractors under any form of contract.	No expiration under current law.	$7.50 billion
Temporary 50 percent expensing for equipment used in the refining of liquid fuels	Taxpayers may elect to expense 50 percent of the cost of qualified refinery property used to process liquid fuel from crude oil and other qualified fuels. The deduction is allowed in the taxable year in which the refinery property is placed in service. The remaining 50 percent of the cost is recovered using a 10-year recovery period. Eligible refineries must have a binding construction contract entered into before January 1, 2010. As of October 3, 2008, qualified refineries include those used in the refining of liquid fuels directly from shale or tar sands.	Eligible refineries must be placed in service before January 1, 2014.	$3.87 billion

Table 2. (Continued)

Name	Description	Expirationinformation	Estimated revenue losses
Exceptions for publicly traded partnership with qualified income derived from certain energy-related activitiesd	The tax code generally treats a publicly traded partnership—i.e., a partnership traded on an established securities market or secondary market—as a corporation for federal income tax purposes. However, a notable exception occurs if 90 percent of the gross income of a partnership is passive-type income, such as interest, dividends, real property rents, gains from the disposition of real property, and similar income or gains. In these cases, the partnership is exempt from corporate level taxation, thus allowing it to claim pass-through status for tax purposes. In general, publicly traded partnerships favor the owners of publicly traded partnerships whose main source of qualifying income is from energy related activities. In contrast to an otherwise similar corporation, the owners of such a publicly traded partnership are not subject to a corporate level tax.	No expiration under current law.	\$3.70 billione
Credit for enhanced oil recovery costs	Taxpayers may claim a credit equal to 15 percent of enhanced oil recovery costs. An enhanced oil recovery project is generally a project that involves the use of one or more tertiary recovery methods to increase the amount of recoverable domestic crude oil. Qualified costs include (1) amounts paid for depreciable tangible property; (2) intangible drilling and development expenses; (3) tertiary injectant expenses; and (4) construction costs for certain Alaskan natural gas treatment facilities. This credit is reduced over a \$6 phase-out range when the reference price for domestic crude oil exceeds \$28 per barrel (adjusted for inflation after 1991). This tax preference is currently phased out due to high crude oil prices.	No expiration under current law.	\$1.98 billion

Name	Description	Expirationinformation	Estimated revenue losses
Capital gains treatment of royalties on coal	Owners of coal mining rights who lease their property usually receive royalties on mined coal. If the owners are individuals, these royalties can be taxed at a lower individual capital gains tax rate rather than at the higher individual top tax rate.	No expiration under current law.	$1.34 billion
Credit for investment in clean coal facilities	An investment tax credit is available for selected types of advanced coal technologies. The Energy Improvement and Extension Act of 2008 allocated $1.25 billion in credits for power generation projects that use integrated gasification combined cycle or other advanced coal-based electricity generation technologies. Qualifying taxpayers may be eligible for a 30 percent credit. The Energy Improvement and Extension Act of 2008 also allocated $250 million in credits for qualified gasification projects (with a credit rate of 30 percent). Prior allocations were awarded under the Energy Policy Act of 2005 and provided $800 million for integrated gasification combined cycle projects and $500 million for other advanced coal-based electricity generation technologies. The Energy Policy Act of 2005 also allocated $350 million for qualified gasification projects.	No expiration under current law (other than the credit allocation limit).	$1.41 billion
Amortization of air pollution control facilities	Prior to 2005, investments in pollution control equipment for pre-1976 coal-fired plants were amortizable over 5 years. In addition, pollution control equipment added to "newer" plants (those placed in service after 1975) was depreciated using the same methods that apply to other electric generating equipment on the date they are placed in service (15- or 20-year recovery period).	No expiration under current law.	$1.30 billione

Table 2. (Continued)

Name	Description	Expirationinformation	Estimated revenue losses
	However, under the Energy Policy Act of 2005, investments in pollution control equipment made in connection with post-1975 power plants qualify for amortization over seven years rather than five years. Qualifying pollution control equipment means any technology that is installed in or on a qualifying facility to reduce air emissions of any pollutant regulated by the Environmental Protection Agency (EPA) under the Clean Air Act.		
Alternative fuel mixture creditf	The tax code provides a 50-cents-per gallon excise tax credit for certain alternative fuels used as fuel in a motor vehicle, motor boat, or airplane and a 50-cents-per gallon credit for alternative fuels mixed with a traditional fuel (gasoline, diesel or kerosene) for use as a fuel. Examples of qualifying fuels include liquefied petroleum gas, compressed or liquefied natural gas, liquefied hydrocarbons derived from biomass, and liquefied hydrogen. If excise tax credits exceeded excise tax liability, the credits could be claimed as income tax credits or received as payments.	Excise tax credits for alternative fuels expired after December 31, 2013. Excise tax credits for liquefied hydrogen fuel will expire after September 30, 2014.	$0.97 billion
Accelerated depreciation recovery periods for specific energy property: natural gasg	A taxpayer is allowed to recover, through annual depreciation deductions, the cost of certain property used in a trade or business or for the production of income. For natural gas distribution lines placed in service between 2005 and 2011, the tax code allows a recovery period over a 15-year period.	Property had to be placed in service before January 1, 2011.	$0.69 billion

Name	Description	Expirationinformation	Estimated revenue losses
Amortization of geological and geophysical expenditures associated with oil and gas exploration	Geological and geophysical costs — exploratory costs associated with determining the precise location and potential size of a mineral deposit — are amortized by independent producers over 2 years and by major integrated oil companies over 7 years.	No expiration under current law.	$0.58 billion
Exception from passive loss limitation for working interests in oil and gas properties	This provision exempts working interests (investments) in gas and oil exploration and development from being categorized as "passive income (or loss)" with respect to the Tax Reform Act of 1986. In general, a working interest is an interest with respect to an oil and gas property that is burdened with the cost of development and operation of the property. The exception allows owners of working interests to offset their losses from passive activities against active income. Under normal rules, passive losses that remain after being netted against passive income can only be carried forward to apply against passive income in future years. The exception from passive loss limitation provision on oil and natural gas properties applies principally to partnerships and individuals rather than corporations. This categorization permits the deduction of losses in oil and gas projects against other active income earned without limitation and is believed to act as an incentive to induce investors to finance oil and gas projects.	No expiration under current law.	$0.31 billion

Table 2. (Continued)

Name	Description	Expirationinformation	Estimated revenue losses
Carbon dioxide sequestration credit	A credit of $10 per metric ton is available for qualified carbon dioxide that is captured by the taxpayer at a qualified facility, used by such taxpayer as a tertiary injectant (including carbon dioxide augmented waterflooding and immiscible carbon dioxide displacement) in a qualified enhanced oil or natural gas recovery project and disposed of by such taxpayer in secure geological storage. In addition, a credit of $20 per metric ton is available for qualified carbon dioxide captured by a taxpayer at a qualified facility and disposed of by such taxpayer in secure geological storage without being used as a tertiary injectant. Both credit amounts are adjusted for inflation after 2009.	No set expiration date. This provision will expire at the end of the year in which the Secretary of Energy determines that 75 million metric tons of carbon dioxide have been captured and sequestered.	$0.22 billion
Deduction for small refiners with capital costs associated with EPA sulfur regulation compliance	A small business refiner may immediately deduct as an expense 75 percent of the costs paid or incurred for purposes of complying with EPA's Highway Diesel Fuel Sulfur Control requirement. A cooperative that qualifies as a small business refiner may elect to pass this deduction through to its owners. Costs qualifying for the deduction are those costs paid or incurred with respect to any facility of a small business refiner during the period beginning on January 1, 2003, and ending on the earlier of the date that is 1 year after the date on which the taxpayer must comply with the applicable EPA regulations or December 31, 2009.	In general, qualifying costs must have been paid or incurred no later than December 31, 2009.	$0.04 billion

Name	Description	Expirationinformation	Estimated revenue losses
Partial expensing of investments in advanced mine safety equipment	Section 404 of the Tax Relief and Welfare Act of 2006 allowed qualified mine safety equipment to be expensed rather than capitalized.	This provision expired as of December 31, 2013.	$0.03 billion

Sources: GAO analysis of Treasury, Joint Committee on Taxation, and Congressional Research Service information. | GAO-14-836

Note: Revenue losses reflect Treasury estimates from the President's budget unless otherwise specified. Revenue loss estimates do not incorporate any behavioral responses and thus do not reflect the exact amount of revenue that would be gained if a specific tax expenditure were repealed. In addition, while sufficiently reliable as a gauge of general magnitude, summing individual tax expenditures' revenue loss estimates does not take into account interactions between individual provisions.

[a]The nonconventional fuel credit may include revenues losses associated with biomass.

[b]Depreciation—a normal business expense under an income tax system—is an annual deduction from income that allows taxpayers to recover the cost or other basis of certain property used in a business or other income-producing activity over the useful life of the property. Depletion, like depreciation, is a form of capital recovery in that an asset (i.e., the mineral reserve itself) is being expended in order to produce income.

[c]Expensing refers to deducting costs in the year they are paid or incurred; capitalizing refers to recovering such costs through depletion or depreciation. Expensing costs is an exception to general tax rules that provide for the capitalization of costs related to generating income from capital assets.

[d]The exception for publicly traded partnerships with qualified income derived from certain energy related activities may include revenue losses not associated with fossil energy production.

[e]The revenue loss estimate is reported by the Joint Committee on Taxation only.

[f]The alternative fuel mixture credit may include revenues losses associated with biomass.

[g]The Joint Committee on Taxation generally classifies as tax expenditures cost recovery allowances that are more favorable than those provided under the alternative depreciation system (Internal Revenue Code Section 168(g)), which provides for straight-line recovery over tax lives that are longer than those permitted under the accelerated system. Accelerated depreciation, in effect, reduces the cost of acquiring energy properties by allowing businesses to deduct larger amounts from their taxable income sooner than they would be able to do under straight-line depreciation. Reducing tax liability earlier provides a benefit to the taxpayer because of the time value of money—having a lower tax payment today is worth more to the taxpayer than having the lower payment in the future.

Table 3. Three Federal Tax Provisions Related to Fossil Energy

Name	Description	Expiration information
Foreign tax credit and dual capacity taxpayers	The tax code provides a tax credit for income taxes paid to foreign countries, which helps to protect taxpayers who earn income abroad from double taxation. If a multinational company is subject to a foreign country's levy, and it also receives a specific economic benefit from that foreign country, it is classified as a "dual-capacity taxpayer." Dual-capacity taxpayers cannot claim a credit for any part of the foreign levy unless it is established that the amount paid under a distinct element of the foreign levy is a tax, rather than a compulsory payment for some direct or indirect economic benefit. JCT estimated that repealing this provision will increase revenue related to oil and gas production by $7.5 billion from fiscal year 2012 through 2022.	No expiration under current law.
Last-in, first-out inventory accounting method	In general, for federal income tax purposes, taxpayers must account for inventories if the production, purchase, or sale of merchandise is a material income-producing factor to the taxpayer. Under the last-in, first-out ("LIFO") method, it is assumed that the last items entered into the inventory are the first items sold. Because the most recently acquired or produced units are deemed to be sold first, cost of goods sold is valued at the most recent costs; the effect of cost fluctuations is reflected in the ending inventory, which is valued at the historical costs rather than the most recent costs. Compared to first-in, first-out ("FIFO"), LIFO produces net income that more closely reflects the difference between sale proceeds and current market cost of inventory. When costs are rising, the LIFO method results in a higher measure of cost of goods sold and, consequently, a lower measure of income when compared to the FIFO method.	No expiration under current law.

Name	Description	Expiration information
	The inflationary gain experienced by the business in its inventory is generally not reflected in income, but rather, remains in ending inventory as a deferred gain until a future period in which sales exceed purchases. JCT estimated that repealing this provision will increase revenue by $106 billion from fiscal year 2014 through 2024; while OMB estimated that repealing this provision will increase revenue by almost $81 billion from fiscal year 2014 through 2023. However, neither JCT nor OMB quantified how much of its revenue estimate is related to fossil fuel production.	
Deduction for income attributable to domestic production activities	Section 199 of the American Jobs Creation Act of 2004 allows a deduction of qualified production activities from taxable income of 3 percent in 2005-2006, 6 percent in 2007-2009, and 9 percent thereafter. The deduction cannot exceed total taxable income of the firm and is limited to 50 percent of wages related to the qualified activity. This provision lowers the effective tax rate on the favored property, in most cases when fully phased in, from the top corporate tax rate of 35 percent to 31.85 percent. Production property is property manufactured, produced, grown, or extracted within the United States. Eligible property also includes domestic film, energy, and construction, and engineering and architectural services. For the latter, the services must be produced in the United States for construction projects located in the United States. The law specifically excludes the sale of food and beverages prepared at a retail establishment, the transmission and distribution of electricity, gas, and water, and receipts from property leased, licensed, or rented to a related party. The benefits are also allowed for Puerto Rico for 2007 through 2011. Oil extraction is permanently limited to a 6 percent deduction. JCT estimated that repealing this provision will increase revenues related to oil and gas production by $14.4 billion from fiscal year 2012 through 2022; while OMB estimated $17.4 billion in revenue from fiscal year 2014 through 2023 related to fossil energy production.	This deduction was allowed for Puerto Rico from 2006 through 2013. Otherwise, there is no expiration under current law.

Sources: GAO analysis of Treasury and Joint Committee on Taxation documentation. | GAO-14-836.

Regarding royalty relief, the federal government provided nearly $12 billion in royalty relief for oil and gas production from 2000 through 2012, according to Interior estimates.[39] As shown in figure 16, revenue losses associated with royalty relief increased from $40 million in 2000 to more than $2 billion in 2011, before declining to about $1.9 billion in 2012 (the most recent estimate available).

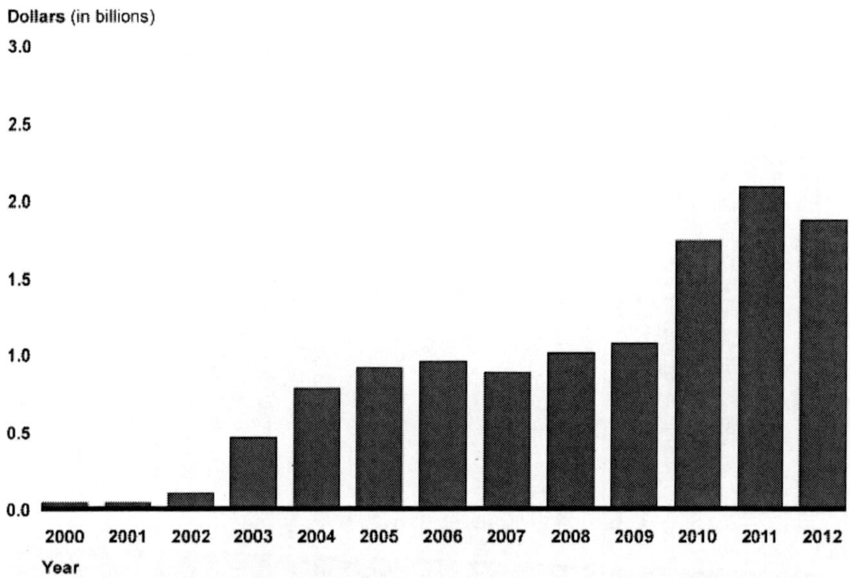

Source: GAO analysis of Interior data. I GAO-14-836.

Note: These estimates reflect forgone revenues in the calendar year in which the sale of product took place irrespective of when the lease was issued. Forgone royalties were calculated by Interior based on production, average annual reported price, and average reported royalty rates in the Gulf of Mexico. 2012 is the latest year for which data on revenue loss estimates were available. Data are not adjusted for inflation.

Figure 16. Estimated Revenue Losses Associated with Royalty Relief for Federal Oil and Gas Leases, 2000-2012.

These federal activities—collecting revenues from excise taxes and royalties and forgoing revenues from tax expenditures and royalty relief—may have influenced U.S. fossil energy production and consumption in different ways, as described below:

- **Excise taxes.** Because excise taxes raised prices on motor fuels, they provided a disincentive for consuming such fuels.[40] However, because much of the revenue from these excise taxes was used to improve roads and other transportation infrastructure, these taxes could also have provided an incentive for motor vehicle use and thereby increased consumption of motor fuels.
- **Royalties.** Because royalty payments raised costs associated with the development and sales of fossil energy, they provided a disincentive to produce and consume fossil energy. However, we cannot say to what extent the federal royalties provided a disincentive for oil and gas development on federal lands relative to other places because oil and gas companies that lease federal lands look for the best economic terms across a wide range of land owners (such as state, private, federal, and international owners). We found in 2008 that studies of many resource owners indicated that the federal government collected less in total revenues than most other resource owners,[41] but we do not have more recent comparisons of revenues collected.
- **Tax expenditures and royalty relief.** In general, tax expenditures and royalty relief provided incentives for fossil energy production by lowering the costs associated with the exploration and development of oil and gas resources.

APPENDIX IV: INFORMATION ON MAJOR FACTORS INFLUENCING NUCLEAR ENERGY PRODUCTION AND CONSUMPTION

This appendix provides more detailed information on U.S. production and consumption of nuclear energy from 2000 through 2013 and on major factors, including federal activities, that may have influenced nuclear energy production and consumption levels.

Nuclear Energy Production and Consumption

The United States has 100 operating commercial nuclear reactors, all of which are used to generate electricity. The year-to-year pattern of domestic production and consumption of nuclear energy fluctuated from 2000 through

2013, as shown in figure 17. In general, nuclear energy production and consumption increased from 754 million megawatt-hours[1] in 2000 to 806 million megawatt-hours in 2007, according to Energy Information Administration (EIA) data. From 2007 through 2010, nuclear energy production and consumption remained steady between about 800 million megawatt-hours and 807 million megawatt-hours. From 2011 through 2013, nuclear energy production and consumption decreased below 800 million megawatt-hours. The proportion of electricity generated by nuclear power in the United States changed little during the period we reviewed, with nuclear reactors accounting for about 20 percent of total U.S. electricity generation in 2000 and about 19 percent in 2013.

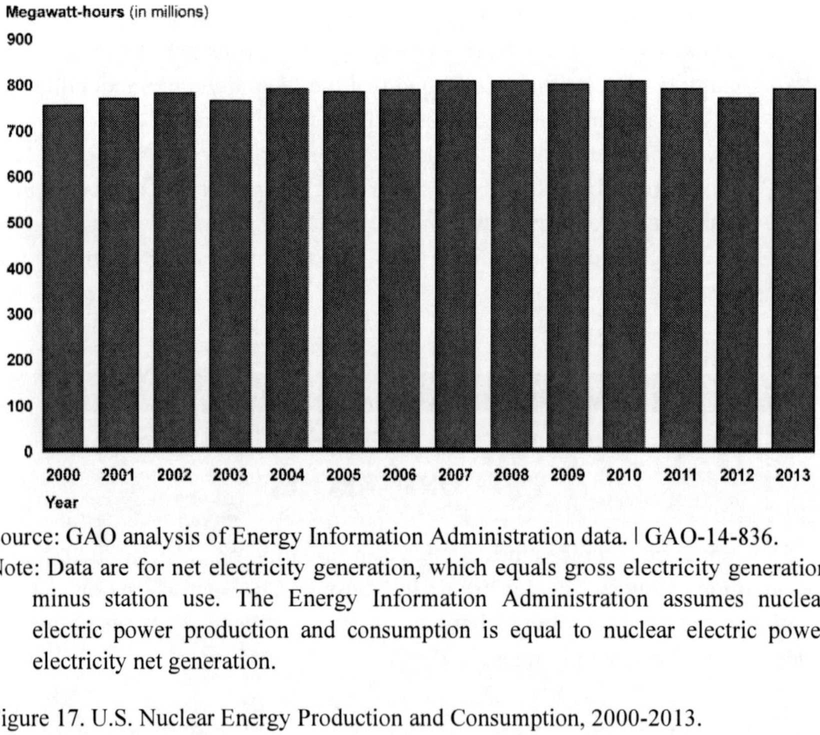

Source: GAO analysis of Energy Information Administration data. | GAO-14-836.

Note: Data are for net electricity generation, which equals gross electricity generation minus station use. The Energy Information Administration assumes nuclear electric power production and consumption is equal to nuclear electric power electricity net generation.

Figure 17. U.S. Nuclear Energy Production and Consumption, 2000-2013.

Major Factors Influencing Nuclear Energy Production and Consumption

The studies and reports we reviewed indicated that recent decreases in U.S. nuclear energy production and consumption may be due to a number of major factors, including reductions in natural gas prices and the Fukushima Daiichi nuclear accident. Regarding natural gas prices, some nuclear plant operators cited price reductions as an important factor in their decisions regarding nuclear power reactor operations. For example, the Vermont Yankee Nuclear Power Station in Vernon, Vermont, began operations in 1972, and the owners obtained a renewed license in 2011 to operate the plant for an additional 20 years. However, in August 2013, the owners announced plans to permanently close the plant in 2014. According to the owners, their decision to close the plant was driven in part by lower natural gas prices, which had reduced the comparative profitability of the plant.

In March 2011, a 9.0-magnitude earthquake and subsequent tsunami devastated northeastern Japan and severely damaged the Fukushima Daiichi nuclear power plant. The resulting radiological emergency involved the most extensive release of radioactive material at a nuclear power plant since the 1986 Chernobyl disaster. Following this release, the Japanese government evacuated people within 12 miles of the plant, and later extended the evacuation zone to 19 miles. In total, almost 150,000 people were evacuated. In response to the incident, Japan shut down all of its nuclear power reactors, and concerns heightened about the safety of commercial nuclear power plants worldwide. For example, Germany closed 8 of the country's 17 reactors and decided to shut down the remainder by 2022. In the United States, the Fukushima incident affected some plans to build new nuclear power plants. For example, in 2011, a company planning to construct two nuclear reactors in Texas cited uncertainties related to the Fukushima incident as a reason for abandoning the project.[2]

U.S. nuclear energy production and consumption trends may also have been affected, to a more limited extent, by increases in the price of uranium oxide, which is processed into fuel used by nuclear power reactors. As shown in figure 18, uranium oxide prices have increased considerably from 2000 to 2013, according to EIA data. Specifically, the average domestic price of uranium oxide increased from $11.45 per pound in 2000 to $52.51 per pound in 2013, an increase of more than 300 percent.

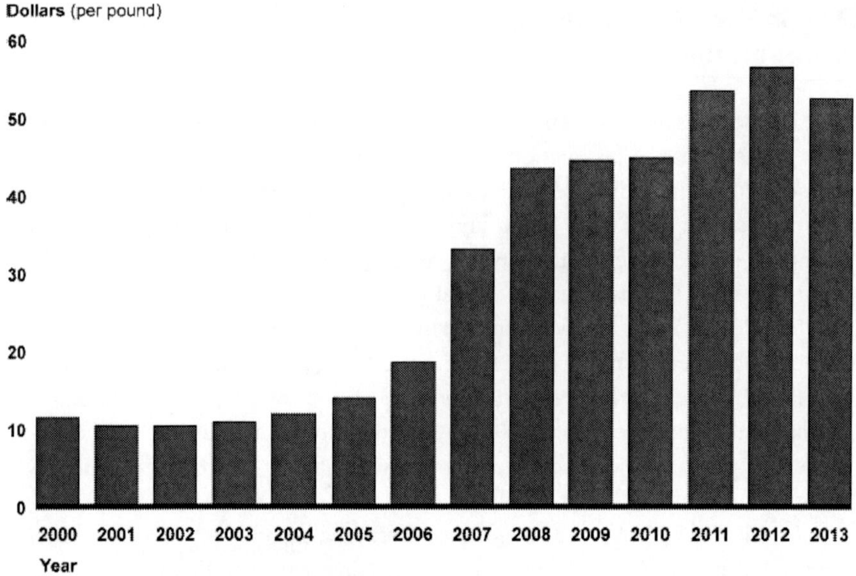

Source: GAO analysis of Energy Information Administration data. | GAO-14-836.

Note: Prices refer to average prices for domestic purchases of uranium oxide. Data are not adjusted for inflation.

Figure 18. U.S. Prices of Uranium Oxide, 2000-2013.

Federal Activities That May Have Influenced Nuclear Energy Production and Consumption

In addition to the major factors we identified above from the studies and reports we reviewed, we identified a number of federal activities that may have also played a role in influencing U.S. production and consumption of nuclear energy from 2000 through 2013. These activities included setting standards and requirements related to the operation of nuclear power reactors, directly providing goods and services related to the storage of spent nuclear fuel, assuming risk associated with the operation of nuclear power reactors, and forgoing revenues associated with tax expenditures for nuclear energy producers. Some of these activities provided an incentive to produce and consume nuclear energy, while others provided a disincentive for its production and consumption.

Setting Standards and Requirements

The federal government established or strengthened a number of standards and requirements related to nuclear energy from 2000 through 2013. For example, after the Fukushima incident, the Nuclear Regulatory Commission (NRC) accepted 12 recommendations from a task force that NRC had convened in 2011 to review its processes and regulations and determine whether lessons learned from the accident could inform its oversight processes. The task force recommended that NRC require licensees to reevaluate and upgrade seismic and flooding protection of reactors and related equipment, strengthen capabilities at all reactors to withstand loss of electrical power, and take other actions to better protect their plants for a low-probability, high-impact event.[3] NRC's activities to strengthen the safety and security of nuclear power plants after the Fukushima incident may have increased the costs associated with operating commercial nuclear power reactors, thereby providing a disincentive for nuclear power production.

Directly Providing Goods and Services

The federal government engaged in activities related to providing for the management of spent nuclear fuel—i.e., fuel that has been used and removed from the reactor core of a nuclear plant—from 2000 through 2013. The nation currently has about 70,000 metric tons of commercial spent nuclear fuel stored at 75 sites in 33 states. This fuel is extremely hazardous: without protective shielding and proper handling, its intense radioactivity can kill a person directly exposed to it or cause long-term health hazards, such as cancer, as well as contaminate the environment. The Nuclear Waste Policy Act of 1982 directed the Department of Energy (DOE) to investigate sites for a federal deep geological repository for spent nuclear fuel and high-level radioactive waste. In 1987, Congress amended the Nuclear Waste Policy Act to direct DOE to focus its efforts only on Yucca Mountain in Nevada for a repository. The act, as amended, authorized DOE to contract with commercial nuclear reactor operators to take custody of spent fuel for disposal at the repository beginning in 1998. DOE spent billions of dollars related to this effort and in 2008 submitted a license application for the construction of a permanent repository at Yucca Mountain to NRC, which has regulatory authority over the construction and operation of a repository.[4]

However, as we reported in August 2012:[5]

- In 2009, DOE announced that it planned to terminate its work related to the Yucca Mountain repository and in 2010 filed a motion to withdraw the license application. NRC's licensing board denied the motion, but DOE continued to take steps to dismantle the repository project.

- In September 2011, the NRC commissioners[6] considered whether to overturn or uphold the licensing board's decision, but they were evenly divided and unable to take final action on the matter. Instead, the NRC commissioners directed the licensing board to suspend work by September 30, 2011.
- NRC's failure to consider the application, among other things, was contested in federal court. Several parties filed a petition against NRC asking the federal court to compel NRC to provide a proposed schedule with milestones and a date for approving or disapproving the license application, among other things.[7]

Federal activities related to the Yucca Mountain repository may have provided a disincentive for nuclear energy production and consumption. For example, DOE's actions regarding its license application for the construction of the repository may have caused uncertainty about the federal government's long-term strategy for storing nuclear waste because Congress has not agreed upon a path forward. This uncertainty may have provided a disincentive for some nuclear plant operators to stay in the market or expand capacity because storing nuclear waste is expensive.

Assuming Risk

The federal government assumed certain risks related to nuclear energy production and consumption from 2000 through 2013. For example, under the Price-Anderson Act, the federal government limited the liability of nuclear plant operators in the case of a nuclear accident.[8] The act requires each licensee of a nuclear plant to have primary insurance coverage equal to the maximum amount of liability insurance available from private sources—currently $375 million—to settle any such claims against it. In the event of an accident at any plant where liability claims exceed the $375 million primary insurance coverage, the act also requires licensees to pay retrospective premiums (also referred to as secondary insurance).[9] The act places a limit on the total liability per incident, which is currently about $13 billion.

In addition, the federal government assumed risks related to nuclear energy production and consumption by establishing a loan guarantee program. Specifically, Section 1703 of the Energy Policy Act of 2005[10] authorized DOE to issue loan guarantees for projects that avoid, reduce, or sequester greenhouse gases using new or significantly improved technologies. In 2010, DOE made conditional commitments under Section 1703 to provide $8.3 billion in loan guarantees for the construction of two advanced nuclear reactors at the Vogtle Electric Generating Plant in Georgia.[11]

These federal activities provided an incentive for nuclear energy production and consumption by decreasing the overall cost associated with certain production-related activities. For example, according to the Congressional Budget Office (CBO), the Price-Anderson Act provides a benefit to nuclear plant operators by reducing their cost of carrying liability insurance.[12] CBO estimated that the potential level of support was about $600,000 annually per reactor, which would be about $62 million annually for all reactors in the United States. Without the liability limitations provided by the Price-Anderson Act, the cost of obtaining insurance for nuclear power plant operators might have been higher. Consequently, the act may have supported higher levels of nuclear power production in the United States between 2000 and 2013 than would have otherwise occurred because the lower cost provided an incentive for increased production and consumption.

Forgoing Revenue

The federal government incurred revenue losses related to nuclear energy production and consumption from 2000 through 2013. Specifically, we identified one tax expenditure targeting nuclear energy that resulted in $7.9 billion in revenue losses from fiscal year 2000 through 2013.[13] This tax expenditure—the special tax rate for nuclear decommissioning reserve funds—increased from $100 million in fiscal year 2000 to $1.1 billion in fiscal year 2013, as shown in figure 19. Under the special tax rate for nuclear decommissioning reserve funds, taxpayers (e.g., utilities) who are responsible for the costs of decommissioning nuclear power plants can elect to create reserve funds to be used to pay for decommissioning. The funds receive special tax treatment: amounts contributed are deductible in the year the contributions are made and are not included in the taxpayer's gross income until the year they are distributed, thus effectively postponing tax on the contributions. Amounts actually spent on decommissioning are deductible in the year they are made. Gains from the funds' investments are subject to a 20 percent tax rate—a lower rate than that which applies to most other corporate income.[14] In general, this tax expenditure supported nuclear energy production and consumption by lowering the costs of nuclear energy production and providing an incentive to engage in nuclear power production.

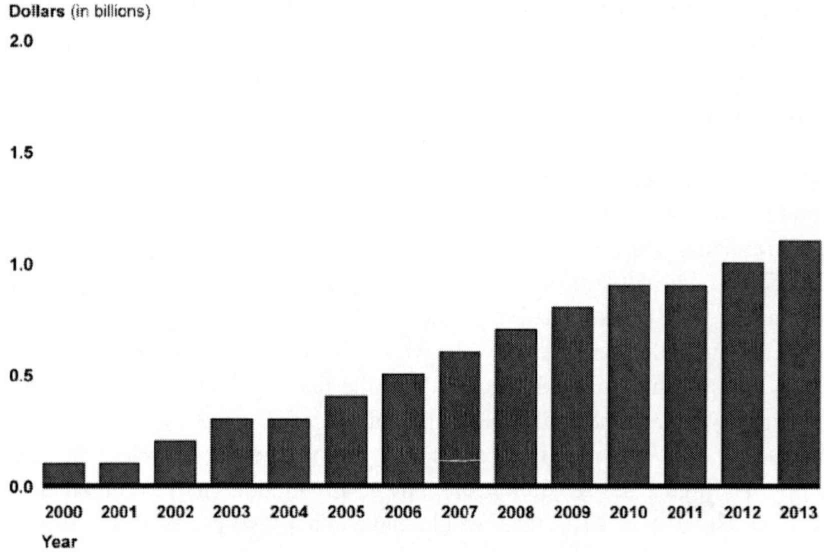

Source: GAO analysis of Joint Committee on Taxation data. | GAO-14-836.

Figure 19. Revenue Losses Associated with a Federal Tax Expenditure Targeting Nuclear Energy Production and Consumption, Fiscal Year 2000 – 2013.

APPENDIX V: INFORMATION ON MAJOR FACTORS INFLUENCING RENEWABLE ENERGY PRODUCTION AND CONSUMPTION

This appendix provides more detailed information on U.S. production and consumption of renewable energy from 2000 through 2013 and on major factors, including federal activities, that influenced renewable energy production and consumption levels.

Renewable Energy Production and Consumption

U.S. production and consumption of renewable energy generally increased from 2000 through 2013, as shown in figure 20, according to Energy Information Administration (EIA) data.[1] Specifically, in terms of energy content (i.e., British thermal units [Btu]), production and consumption of all sources of renewable energy—including liquid biofuels (such as ethanol and biodiesel), other forms of biomass (i.e., wood and waste), hydroelectric power, geothermal, wind, and solar—increased from 6.1 quadrillion Btus in 2000 to 9.2 quadrillion Btus in 2013. As a proportion of total energy consumption, consumption of renewable energy increased from about 6 percent in 2000 to 9 percent in 2013. Hydroelectric power and wood were the two largest sources of renewable energy during this time period and accounted for about 50 percent of all renewable energy consumed in 2013. However, the overall increase in domestic production and consumption of renewable energy from 2000 through 2013 can be attributed primarily to increases in the production and consumption of ethanol, wind energy, and solar energy.

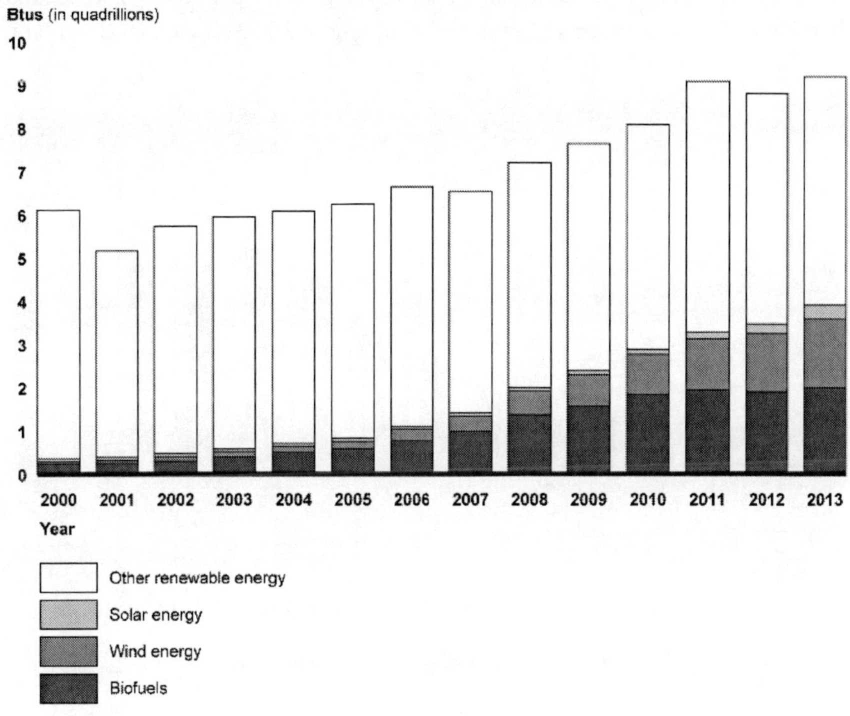

Source: GAO analysis of Energy Information Administration data. | GAO-14-836.

Note: Btu stands for British thermal unit. Other renewable energy includes energy from forms of biomass (i.e., wood, waste) other than biofuels, hydroelectric power, and geothermal energy. Production equals consumption for all renewable energy sources except biofuels.

Figure 20. U.S. Renewable Energy Production and Consumption, 2000-2013.

Ethanol Production and Consumption

Domestically produced ethanol primarily comes from corn grown in the Midwest; the cornstarch is converted into sugar and then fermented and distilled into ethanol. Ethanol is used as a transportation fuel; almost all ethanol is blended into gasoline as an additive to make fuel containing up to 10 percent ethanol by volume.[2] As shown in figure 21, ethanol production increased from 1.6 billion gallons in 2000 to almost 14 billion gallons in 2011 and about 13 billion gallons in both 2012 and 2013. Consumption of ethanol followed a similar pattern until 2010, when domestic consumption of ethanol remained relatively flat at around 13 billion gallons per year. The United States increased its exports of ethanol beginning in 2010, mostly to Brazil and Canada.[3]

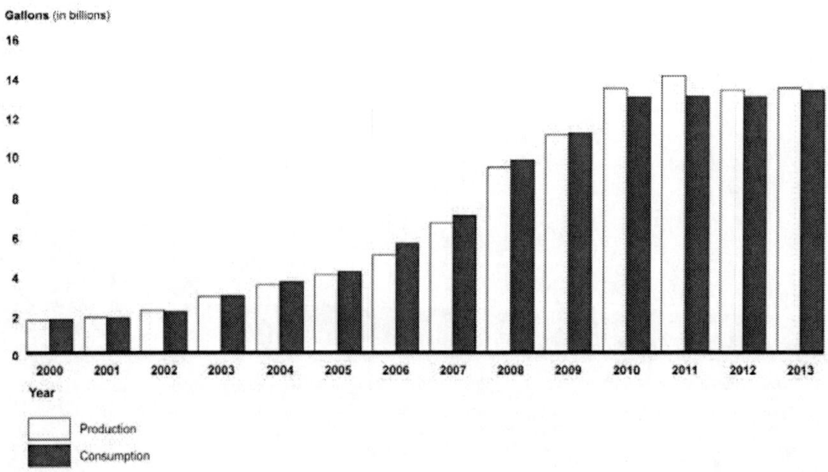

Source: GAO analysis of Energy Information Administration data. | GAO-14-836.

Figure 21. U.S. Ethanol Production and Consumption, 2000-2013.

Major Factors Influencing Ethanol Production and Consumption

The studies and reports we reviewed indicated that several federal activities had a major impact on the increase in ethanol production and consumption—most notably federal tax expenditures and requirements for the use of ethanol in transportation fuel. Regarding federal tax expenditures, alcohol fuel credits provided a 45-cent-per-gallon tax credit to gasoline suppliers who blend ethanol with gasoline.[4] According to the Department of the Treasury (Treasury) data, the alcohol fuel credits resulted in more than $39 billion in revenue losses from fiscal year 2000 through 2013. As shown in figure 22, revenue losses associated with alcohol fuel credits increased from about $0.9 billion in fiscal year 2000 to $7 billion in fiscal year 2011, before decreasing to $3.7 billion in fiscal year 2012 and about $50 million in fiscal year 2013.[5]

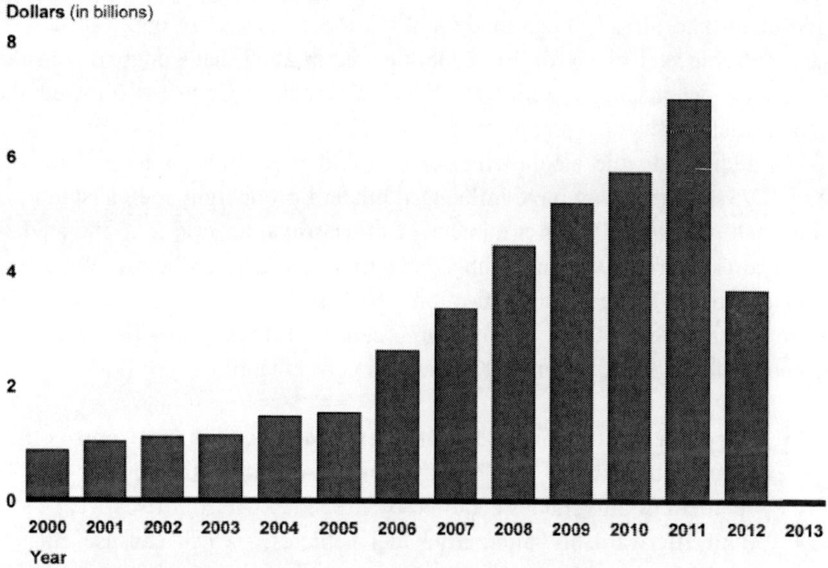

Source: GAO analysis of Office of Management and Budget and Treasury data. | GAO-14-836.

Note: Data are not adjusted for inflation.

Figure 22. Revenue Losses Associated with Alcohol Fuel Credits, Fiscal Year 2000 – 2013.

Regarding federal requirements, the Energy Policy Act of 2005 (EPAct) created a federal renewable fuel standard that generally required gasoline and diesel sold in the United States to contain 4 billion gallons of renewable fuels, such as ethanol and biodiesel, in 2006 and 7.5 billion gallons in 2012.[6] The Energy Independence and Security Act of 2007 (EISA) expanded the renewable fuel standard by requiring that U.S. transportation fuel contain 9 billion gallons of renewable fuels in 2008, with the amount required increasing annually to 36 billion gallons in 2022.[7] The 36-billion-gallon total must include at least 21 billion gallons of advanced biofuels (defined as renewable fuels other than ethanol derived from corn starch that meet certain criteria) and can include up to 15 billion gallons of conventional biofuels (defined as ethanol derived from cornstarch).

In our previous work, we found that the alcohol fuel credits were important in establishing and expanding the domestic ethanol industry.[8] However, we also found that the alcohol fuel credits became less important over time for sustaining the ethanol industry because (1) most of the capital investment had already been made and (2) the credits were duplicative with the renewable fuel standard. We recommended in 2009 that Congress consider modifying or phasing out the alcohol fuel credits.[9] Congress allowed the alcohol fuel credits to expire at the end of 2011.[10]

In addition to the alcohol fuel credits and renewable fuel standard, the federal government may have influenced ethanol production and consumption from 2000 through 2013 in a number of other ways, according to the studies and reports we reviewed, including the activities described below. While the precise effects of these activities on changes in ethanol production and consumption are difficult to measure, some of these activities provided incentives for the production and consumption of ethanol.

- **Requirements for federal fleets to use ethanol and other alternative fuels.** The Energy Policy Act of 1992 requires that 75 percent of all vehicles acquired by the federal fleet in fiscal year 1999 and afterward be "alternative fuel vehicles," which can use ethanol and blends of 85 percent or more of ethanol with gasoline, among other fuels. EPAct generally requires that all such vehicles be fueled with alternative fuel.[11] In addition, EISA[12] requires that no later than October 2015 and each year thereafter, agencies must achieve a 10 percent increase in vehicle alternative fuel consumption relative to a baseline established by the Energy Secretary for fiscal year 2005.
- **Excise taxes.** The federal excise tax rate on ethanol in motor fuels is 18.4 cents per gallon.[13] We did not separately analyze the portion of revenues from excise taxes on ethanol from the portion on gasoline. However, we believe the effects of these excise taxes may be similar to the effects of excise taxes on gasoline.

Another factor likely affecting ethanol production and consumption from 2000 through 2013 was the price of ethanol relative to the prices of corn and gasoline, according to U.S. Department of Agriculture (USDA) research. Ethanol prices generally increased from 2000 through 2013, according to USDA data, as shown in figure 23. Specifically, ethanol prices increased from an annual average of $1.35 per gallon in 2000 to $2.47 per gallon in 2013. Because ethanol is used as a gasoline substitute, and because nearly all ethanol produced in the United States comes from corn, the relationship between prices of ethanol, gasoline, and corn is complex. As gasoline prices rise, ethanol's appeal as a substitute increases, as does the profitability of ethanol production and the demand for corn. As a result, according to USDA's Economic Research Service, prices of corn, ethanol, and gasoline have become more interrelated in recent years.[14] Specifically, from March 2008 to March 2011, ethanol supply and demand accounted for about 23 percent of the variation in the price of corn, while corn market conditions accounted for about 27 percent of ethanol's price variation. At the same time, about 16 and 17 percent of gasoline price variation could be attributed to ethanol and corn markets conditions, respectively.

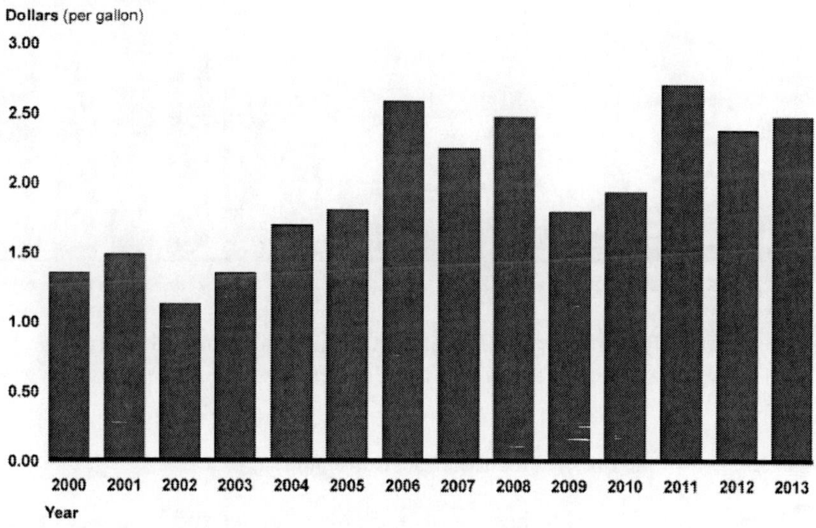

Source: GAO analysis of USDA Economic Research Service data. | GAO-14-836.

Note: Prices are rack prices (wholesale truckload sales or smaller of gasoline where title transfers at a terminal). Data are not adjusted for inflation.

Figure 23. U.S. Ethanol Prices, 2000-2013.

Production and Consumption of Wind and Solar Energy

Wind is transformed into electricity using wind turbines. In terms of the electricity generated from wind turbines, domestic production and consumption of wind energy increased from 5.6 million megawatt-hours in 2000 to 167.7 million megawatt-hours in 2013 (an increase of almost 3,000 percent), as shown in figure 24. According to the Department of Energy (DOE), wind energy comprised 43 percent of all additions to U.S. generating capacity in 2012, overtaking natural gas-fired electricity generation as the leading source of new capacity for that year.[15]

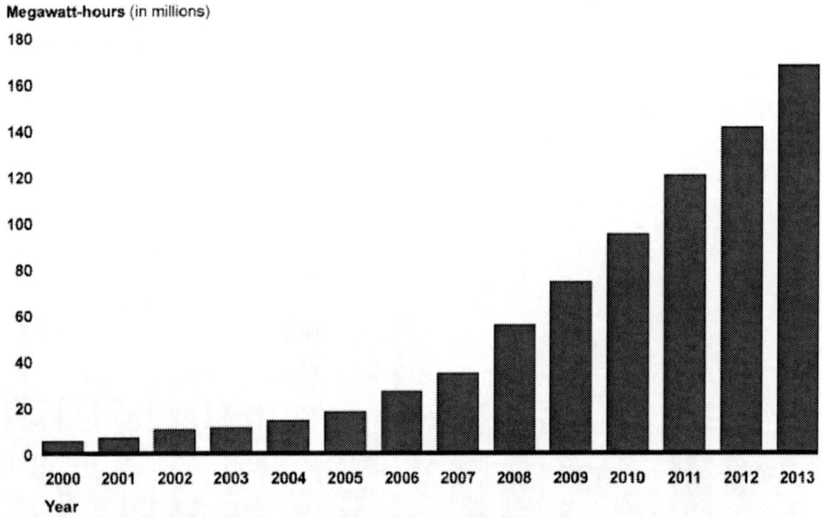

Source: GAO analysis of Energy Information Administration data. | GAO-14-836.

Note: Data are for net electricity generation. Production equals consumption for wind energy.

Figure 24. U.S. Wind Energy Production and Consumption, 2000-2013.

Solar energy is used to heat, cool, and power homes and businesses using a variety of technologies that convert sunlight into usable energy. The most widely used solar technology is the photovoltaic cell, which uses semiconducting materials to convert sunlight into electricity. Sunlight can also be used to heat water (for later use) or to boil water, which produces steam that can be used in a turbine to generate electricity. In terms of the electricity generated from solar energy, domestic production and consumption of solar energy increased from about 0.5 million megawatt-hours in 2000 to 9.3 million megawatt-hours in 2013 (an increase of almost 1,900 percent), as shown in figure 25. According to the Solar Energy Industries Association, photovoltaic solar installations grew 76 percent from 2011 to 2012, and 8 of the 10 largest photovoltaic installations in the United States were built in 2012.[16]

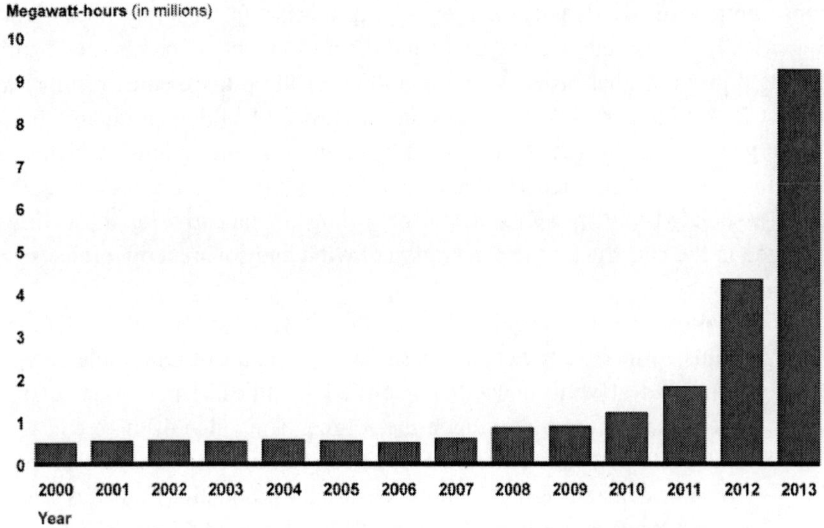

Source: GAO analysis of Energy Information Administration data. | GAO-14-836.
Note: Data are for net electricity generation. Production equals consumption for solar energy.

Figure 25. U.S. Solar Energy Production and Consumption, 2000-2013.

Major Factors Influencing Production and Consumption of Wind and Solar Energy

The studies and reports we reviewed indicated that the increase in wind and solar energy production and consumption resulted from a number of major factors—most notably state policies and federal activities, as well as technological advances. Regarding state activities, many states have created policies known as renewable portfolio standards that encouraged the production and use of renewable energy. These state policies generally require a percentage of electricity sold or generated in the state to come from eligible renewable resources, including wind and solar energy. According to EIA, 29 states and the District of Columbia had enforceable renewable portfolio standards or similar laws as of October 2013.[17] According to the Congressional Research Service (CRS), state policies have been the primary creator of demand for wind projects.[18]

Regarding federal activities, the studies and reports we reviewed indicated that the federal government influenced increases in the production and consumption of wind and solar energy primarily through tax incentives. Specifically, the production tax credit and the investment tax credit, along with a related program that provided grants in lieu of these tax credits, resulted in almost $14 billion in revenue losses and almost $20 billion in outlays from fiscal year 2000 through 2013. These tax credits and grants, which are described below, supported wind and solar energy production by lowering the costs associated with production and providing an incentive to those firms engaged in the construction and operation of wind and solar energy projects.

- **Production tax credit.** This credit provided a 10-year, inflation-adjusted income tax credit based on the amount of renewable energy produced at wind and other qualified facilities. The amount of the credit varied depending upon the source. The value of the credit was 2.2 cents per kilowatthour in 2012 for certain resources (e.g., wind, geothermal, and certain biomass electricity production) and was raised to 2.3 cents per kilowatthour in 2013. This credit resulted in about $9.6 billion in revenue losses from fiscal year 2000 through 2013. Specifically, as shown in figure 26, revenue losses associated with this tax credit increased from $40 million in fiscal year 2000 to $1.5 billion or more annually from fiscal year 2010 through 2013. This credit, which has periodically expired and then been extended, is available to facilities for which construction began before January 1, 2014.[19] As we reported in March 2013, new additions of wind energy capacity fell dramatically in years following the credit's expiration.[20]
- **Investment tax credit.** This credit, which has not expired, provides an income tax credit for business investments in solar systems and small wind turbines, among other things.[21] Investments in solar and small wind turbine systems qualify for a 30 percent tax credit. In addition, temporary provisions enacted under the American Recovery and Reinvestment Act of 2009 (Recovery Act) allow taxpayers to claim this credit for property that otherwise would have qualified for the production tax credit. This credit resulted in over $4 billion in

revenue losses from fiscal year 2000 through 2013. As shown in figure 26, no revenue loss estimates were reported for this tax credit from fiscal year 2000 through 2005; revenue losses then generally increased from $80 million in fiscal year 2006 to almost $2 billion in fiscal year 2013.[22]

- **Section 1603 program.** Section 1603 of the Recovery Act, as amended, allows taxpayers eligible for the production or investment tax credit to receive a payment from the Treasury in lieu of a tax credit. This Treasury program provided almost $20 billion in outlays from fiscal year 2009 through 2013, as shown in figure 26, of which about $13 billion were related to wind energy projects, and about $4 billion were associated with solar energy projects. This program, which is still available in some cases, applies to projects placed in service during 2009, 2010, or 2011, or afterward if construction began on the property during the specific years and the property is placed in service by a credit termination date (e.g., January 1, 2017 for certain energy property).

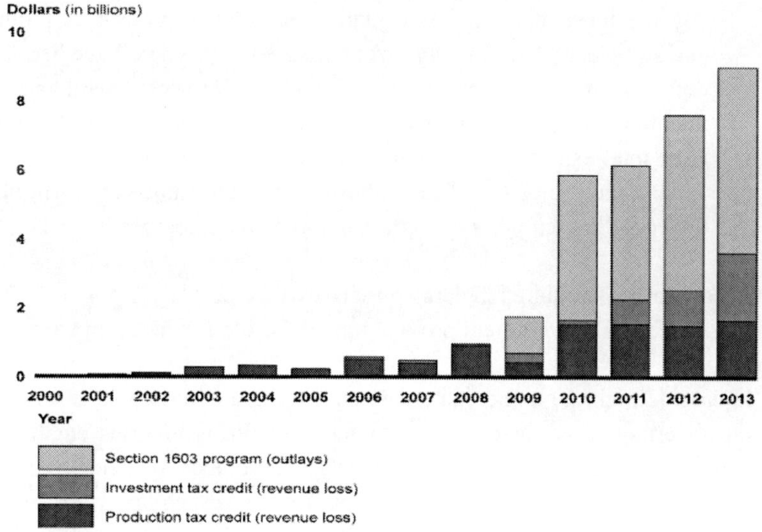

Source: GAO analysis of Office of Management and Budget and Treasury data. | GAO-14-836.

Note: Data are not adjusted for inflation.

Figure 26. Revenue Losses and Outlays Associated with the Production Tax Credit, Investment Tax Credit, and Section 1603 Program, Fiscal Year 2000 – 2013.

In addition to these tax credits, the studies and reports we reviewed indicated that the federal government provided incentives for the production and consumption of wind and solar energy in other important ways, including through the following activities:

- **Requirements for purchasing electricity.** Under EPAct, federal agencies' consumption of electricity from renewable sources has generally been required— to the extent economically feasible and technologically practicable—to meet or exceed 5 percent of total consumption in fiscal years 2010 through 2012, and 7.5 percent in fiscal year 2013 and thereafter.[23] According to DOE's most recent data, federal agencies spent about $57 million in electricity purchases from renewable sources in fiscal year 2012.
- **Loan guarantees.** DOE's Title 17 Innovative Technology Loan Guarantee Program included a temporary program for the rapid deployment of renewable energy projects, among other things. As shown in table 4, DOE guaranteed 23 loans totaling more than $14 billion for wind and solar energy projects. Most of these loans (15 of 23) and most of the amount guaranteed went to projects to produce and sell electricity generated from solar energy. There have been two defaults on guaranteed loans, both for projects involving the manufacture of solar energy equipment. However, most of the long-term total estimated cost to the government is associated with solar generation projects.[24] The authority to enter into loan guarantees under DOE's temporary program expired on September 30, 2011.

As a result of required federal purchases of electricity from renewable sources, the federal government provided incentives to produce wind and solar energy.

In addition, through the loan guarantee program described above, the federal government assumed risks of defaults on loans to firms engaged in developing wind and solar energy projects. These federal actions had the potential to lower the costs for some of these projects. Such lower costs could have led to certain projects being financed that otherwise may not have been developed.

Table 4. DOE's Title 17 Innovative Technology Loan Guarantee Program Targeting Solar and Wind Energy, Fiscal Year 2000 – 2013

Dollars in billions				
Type	**Number of loan guarantees and loans[a]**	**Amount of loan guarantees and loans[b]**	**Number of defaults**	**Estimated cost to the government[c]**
Solar generation	15	$11.62	0	$1.03
Solar manufacturing	4	$1.23	2	$0.62
Wind generation	4	$1.70	0	$0.04
Total	**23**	**$14.55**	**2**	**$1.69**

Sources: GAO analysis of DOE and Office of Management and Budget data. | GAO-14-836

[a]The number of guarantees and loans refers to all guarantees and loans that were issued, including three that were withdrawn or deobligated before any funds were drawn on the loans.

[b]The loan guarantee and loan amounts are the amounts at closing that appear in DOE's accounting system. They include the full amount of the loans partially guaranteed through the Financial Institution Partnership Program and do not include capitalized interest.

[c]These costs are current estimates of the credit subsidy costs of disbursed amounts as reported in the President's fiscal year 2015 budget. Credit subsidy costs represent the government's estimated net long-term cost of extending or guaranteeing credit, in present value terms, over the entire period the loans are outstanding (not including administrative costs).

Along with federal activities, the studies and reports we reviewed indicated that technological advances played a role in influencing increases in production and consumption of wind and solar energy.

For example, according to DOE's National Renewable Energy Laboratory (NREL), wind turbine manufacturers increased turbine performance by steadily increasing the turbine height and rotor diameter of their turbines from 2000 through 2010.

In addition, the average capacity of wind turbines installed in the United States has more than doubled since 2000—increasing from 0.88 megawatts in 2000 to 1.95 megawatts in 2012, according to NREL.

Regarding solar technology, technological innovation—along with improved manufacturing processes and growing markets—resulted in declining costs associated with the manufacture of photovoltaic technologies, according to a 2012 DOE study.[25]

APPENDIX VI: FEDERAL TAX EXPENDITURES, OUTLAYS, AND LOAN GUARANTEES RELATED TO RENEWABLE ENERGY

Table 5 provides descriptions of the three federal tax expenditures we identified in appendix V as targeting or related to ethanol, wind energy, and solar energy, as well as four additional federal tax expenditures we identified as more broadly targeting or related to renewable energy.

The table also provides information from the Department of the Treasury (Treasury) on tax expenditures that will or have expired, in full or in part, due to an expiration of legislative authority or some other expiration under the law as of the fall of 2014, as well as on tax expenditures that currently have no expiration. In addition, the table provides information on revenue loss estimates from Treasury (unless otherwise specified).

Table 6 provides a description of the federal program we identified in appendix V as targeting or related to wind and solar energy, as well as two additional federal programs we identified as more broadly targeting or related to renewable energy.

The table also provides information reported by the Office of Management and Budget (OMB) on outlays.

Table 7 provides a description of the federal loan guarantee program we identified in appendix V as targeting or related to wind and solar energy, as well as an additional federal loan guarantee program we identified as more broadly targeting or related to renewable energy.

The table also provides information on disbursements and estimated costs reported by OMB and provided by DOE.

Table 5. Seven Federal Tax Expenditures Targeting or Related to Renewable Energy, Fiscal Year 2000 – 2013

Name	Description	Expiration information	Estimated revenue losses
Alcohol fuel credits (including related excise tax credit)	The tax code provides three income tax credits for alcohol-based motor fuels: the alcohol mixture credit (or blender's credit), the pure alcohol fuel credit, and the small ethanol producer credit. The alcohol mixture credit is 45¢ per gallon of ethanol of at least 190 proof and is available to the blender (e.g., the refiner, wholesale distributor, or marketer). The alcohol mixture credit is typically claimed as an instant excise tax credit (referred to as the volumetric ethanol excise tax credit). The pure alcohol fuel credit is 45¢ per gallon of ethanol of at least 190 proof and can only be claimed by the consumer or retail seller. For small ethanol producers, the law also provides for a production tax credit in the amount of 10¢ per gallon of ethanol produced and sold for use as a transportation fuel. This credit is limited to the first 15 million gallons of annual alcohol production for each small producer, defined as one with an annual production capacity of fewer than 60 million gallons. This is in addition to any blender's tax credit claimed on the same fuel. In addition, the tax code provides an income tax credit for cellulosic biofuels. The amount of the credit is $1.01 per gallon. In the case of cellulosic biofuel that is alcohol, the credit amount is reduced.	Most of these provisions expired as of December 31, 2011; the tax credit for cellulosic biofuels expired s of December 31, 2013.	$39.28 billion
Production tax credit (also called energy production credit)a	Taxpayers producing energy from a qualified renewable energy source may qualify for a tax credit on a per-kilowatt-hour basis. Qualified energy sources include wind, solar energy, geothermal energy, closed-loop and open-loop biomass, small irrigation power, municipal solid waste, qualified hydropower production, and marine and hydrokinetic renewable energy sources.	Construction must have begun before January 1, 2014.	$9.59 billion

Table 5. (Continued)

Name	Description	Expiration information	Estimated revenue losses
	The credit amount in 2012 was 2.2 cents per kilowatt-hour for wind, solar, closed-loop biomass, and geothermal energy sources and 1.1 cents per kilowatt-hour for other energy sources. The credit amount is based on the 1993 value of 1.5 cents per kilowatt-hour, which is adjusted annually for inflation. This credit is generally available for 10 years, beginning on the date when the facility is placed in service. For facilities placed in service during 2009, 2010, and 2011, taxpayers could claim an investment tax credit or Section 1603 cash payment in lieu of receiving the production tax credit.		
Biodiesel and small agri-biodiesel producer tax credits (including related excise tax credit)	The tax code provides three income tax credits for biodiesel: the biodiesel fuel mixtures credit (i.e., blends of biodiesel and petroleum diesel); the unblended (pure) biodiesel credit, which is either used or sold at retail by the taxpayer; and the small biodiesel producer credit. These tax credits are $1.00 per gallon of biodiesel, including agri-biodiesel (i.e., biodiesel made from virgin oils) and renewable biodiesel. The mixtures tax credit may be claimed as an instant excise tax credit against the 24.4 cents per gallon tax on diesel blends. In addition, the tax code provides an income tax credit of 10 cents per gallon for the first 15 million gallons of agri-biodiesel produced by small agri-biodiesel producers each year. Small agri-biodiesel producers are defined as those with a production capacity less than 60 million gallons per year. This credit can be taken in addition to the $1.00 per gallon income or excise tax credit on the sale of the agri-biodiesel produced by small producers.	These provisions expired as of December 31, 2013.	$5.84 billion

Name	Description	Expiration information	Estimated revenue losses
Investment tax credit (also called energy investment credit)b	The tax code provides an income tax credit for business investments in solar, fuel cells, small wind turbines, geothermal systems, microturbines, and combined heat and power. Solar, fuel cell, and small wind turbine investments qualify for a 30 percent credit. (The credit for fuel cells is limited to $1,500 per 0.5 kilowatt of capacity.) The tax credit for investments in geothermal systems, microturbines, and combined heat and power is 10 percent. (The credit for microturbines is limited to $200 per kilowatt of capacity.) Provisions enacted as part of the American Recovery and Reinvestment Act of 2009 (Recovery Act) allow (1) taxpayers to elect to claim this credit for property that otherwise would have qualified for the production tax credit and (2) taxpayers eligible for this credit to receive a Section 1603 payment from the Treasury in lieu of tax credits.	In general, this provision will expire on December 31, 2016; however, the credit for solar investments will decrease to 10 percent, and the credit for geothermal investments will remain at 10 percent.	$4.30 billion
Accelerated depreciation recovery periods for specific energy property: renewable energyc	A taxpayer is allowed to recover, through annual depreciation deductions, the cost of certain property used in a trade or business or for the production of income. The tax code provides a 5-year recovery period for certain renewable energy equipment, including solar, wind, geothermal, fuel cell, combined heat and power, and microturbine property. Renewable energy generation property that is part of a "small electric power facility" and certain biomass property are also recovered over 5 years. However, the Economic Stimulus Act of 2008 included a 50 percent first-year bonus depreciation provision for a wide range of eligible properties including renewable energy systems. This provision was extended by the Recovery Act, and by the Creating Small Business Jobs Act of 2010.	Property had to be placed in service by January 1, 2014 (or January 1, 2015 for certain other assets) to quality for bonus depreciation.	$1.70 billiond

Table 5. (Continued)

Name	Description	Expiration information	Estimated revenue losses
	Bonus depreciation was further extended through 2012 by the Tax Relief, Unemployment Insurance Reauthorization, and Job Creation Act of 2010, with a 100 percent deduction allowed for property acquired after September 8, 2010, and before January 1, 2012. The American Taxpayer Relief Act of 2012 extended 50 percent expensing for qualifying property purchased and placed in service before January 1, 2014. The 50 percent bonus depreciation narrowed any tax differences between eligible assets based on cost recovery provisions, and the 100 percent bonus depreciation eliminated those differences altogether under the provision for allowing a full write-off of asset acquisition costs.	The 5-year recovery period for certain solar equipment will expire on December 31, 2016.	
Advanced energy property credite	The Recovery Act established a 30 percent tax credit for qualified investments in advanced energy property. Advanced energy projects that may qualify for the tax credit include those that reequip, expand, or establish eligible manufacturing facilities. Facilities that produce the following types of property may qualify: (1) property designed to produce energy using a renewable resource (i.e., solar, wind, geothermal), (2) fuel cells, microturbines, or energy storage systems for use with electric or hybrid-electric vehicles, (3) advanced transmission technologies that support renewable generation (including storage), (4) carbon capture and sequestration property, (5) property designed to refine or blend renewable fuels, (6) energy conservation technologies (i.e., energy-saving lighting or smart grid technologies), (7) plug-in electric vehicles and components, and (8) other advanced energy property designed to reduce greenhouse gas emissions.	No expiration under current law (other than the credit allocation limit).	$1.40 billion

Name	Description	Expiration information	Estimated revenue losses
	A total of $2.3 billion was allocated for advanced energy property investment tax credits. The tax credits were competitively awarded by the Department of Energy (DOE) and Treasury. Taxpayers receiving this credit cannot also claim the investment tax credit.		
Credit for holding clean renewable energy bonds	New clean renewable energy bonds help tax-exempt entities finance capital expenditures for new facilities that produce electricity from renewable sources. Bond holders receive tax credits at 70 percent of the tax credit interest rate, in lieu of interest payments. These bonds may be issued by a public power provider, a cooperative electric company, a governmental body, a clean renewable energy bond lender or a not-for-profit electric utility that has received a loan or loan guarantee under the Rural Electrification Act. Treasury publicly solicited applications for an initial volume cap, set by Congress at $800 million, and awarded allocations based on criteria and applications received. An additional $1.6 billion in new clean renewable energy bond authorization was provided under the Recovery Act. In March 2010, provisions included in the Hiring Incentives to Restore Employment Act allowed issuers of clean renewable energy bonds (and other qualified tax-credit bonds) to receive a direct payment from the Treasury instead of providing tax credits to bondholders.	No expiration under current law.	$0.43 billion

Sources: GAO analysis of Treasury, Joint Committee on Taxation, and Congressional Research Service information. | GAO-14-836

Note: Revenue losses reflect Treasury estimates from the President's budget unless otherwise specified. Revenue loss estimates do not incorporate any behavioral responses and thus do not reflect the exact amount of revenue that would be gained if a specific tax expenditure was repealed. In addition, while sufficiently reliable as a gauge of general magnitude, summing individual tax expenditures' revenue loss estimates does not take into account interactions between individual provisions.

[a]The production tax credit includes revenue losses from the new technology credit.

[b]The investment tax credit may include revenue losses associated with combined heat and power, fuel cells, and microturbines.

[c]The accelerated depreciation recovery periods for specific energy property: renewable energy may include revenue losses associated with combined heat and power and microturbines. The Joint Committee on Taxation generally classifies as tax expenditures cost recovery allowances that are more favorable than those provided under the alternative depreciation system (Internal Revenue Code Section 168(g)), which provides for straight-line recovery over tax lives that are longer than those permitted under the accelerated system. Accelerated depreciation, in effect, reduces the cost of acquiring energy properties by allowing businesses to deduct larger amounts from their taxable income sooner than they would be able to do under straight-line depreciation. Reducing tax liability earlier provides a benefit to the taxpayer because of the time value of money—having a lower tax payment today is worth more to the taxpayer than having the lower payment in the future.

[d]The revenue loss estimate is reported by the Joint Committee on Taxation only.

[e]The advanced energy property credit may include revenue losses associated with fuel cells, microturbines, or energy storage systems for use with electric or hybrid-electric vehicles.

Table 6. Three Federal Programs with Outlays Targeting or Related to Renewable Energy, Fiscal Year 2000 – 2013

Name	Description	Outlays
Payments for specified energy property in lieu of tax credits, Department of the Treasury	Section 1603 of the Recovery Act, as amended, established a program to provide payments to eligible applicants who place specified energy property (related to renewable energy, among other things) in service for use in a trade or business. Applicants could take the payment in lieu of either a production or investment tax credit. These payments provide an incentive for investment in property for electricity production, particularly those applicants without sufficient tax liability to utilize a nonrefundable tax credit. The program provided payments for eligible energy projects placed in service during 2009, 2010, or 2011, or after 2011 if construction began on the property during 2009, 2010, or 2011 and the property is placed in service by a certain date known as the credit termination date (e.g., Jan. 1, 2017 for certain energy property).	$19.65 billiona

Name	Description	Outlays
Bioenergy for Advanced Biofuels program and Repowering Assistance program, U.S. Department of Agriculture	The Bioenergy for Advanced Biofuels program, authorized under the Food, Conservation, and Energy Act of 2008, provides payments to eligible producers to support and ensure an expanding production of advanced biofuels, which are defined as fuel derived from renewable biomass other than corn kernel starch. The amount of each payment depends on the number of producers participating in the program, the amount of advanced biofuels being produced, and the amount of funds available. The Repowering Assistance program, authorized under the Food, Conservation, and Energy Act of 2008, provides payments to biorefineries to replace fossil fuels with renewable biomass as a means to produce heat and power. To be eligible, the biorefineries must be located in a rural area and must have been in existence as of June 18, 2008. Payments are available for periods of up to 3 years.	$0.24 billion
Payment in lieu of tax credit for holders of clean renewable energy bonds, Department of the Treasury	Issuers of clean renewable energy bonds can choose to receive a direct payment from the federal government in lieu of the tax credit for bondholders.	$0.06 billion

Sources: GAO analysis of Office of Management and Budget, Energy Information Administration, and Congressional Research Service documentation. | GAO-14-836

[a]Section 1603 payments include payments associated with combined heat and power, fuel cells, microturbines, and other energy technologies.

Table 7. Two Federal Loan Guarantee Programs Targeting or Related to Renewable Energy, Fiscal Year 2000 – 2013

Name	Description	Disbursements	Estimated costsa
Title 17 Innovative Technology Loan Guarantee Program, DOE	This program, originally authorized by Section 1703 of the Energy Policy Act of 2005, was amended by Section 406 of the Recovery Act to establish a temporary program for the rapid deployment of renewable energy and electric power transmission projects, as well as leading-edge biofuels projects.	$15.33 billionb	$1.72 billionc
Biorefinery Assistance Program, U.S. Department of Agriculture	This program, authorized under the Food, Conservation, and Energy Act of 2008, assists in the development of new and emerging technologies for producing advanced biofuels through loan guarantees to fund the development, construction, and retrofitting of commercial-scale biorefineries using an eligible technology. Eligible technologies include those that are being adopted or have been demonstrated to have the potential for application in a commercial-scale biorefinery that produces an advanced biofuel. The maximum available loan guarantee under the program is $250 million.	$0.22 billion	$0.04 billion

Sources: GAO analysis of Office of Management and Budget, Department of Energy, and Energy Information Administration documentation. | GAO-14-836

[a]These costs are current estimates of the credit subsidy costs of disbursed amounts. Credit subsidy costs represent the government's estimated net long-term cost of extending or guaranteeing credit, in present value terms, over the entire period the loans are outstanding (not including administrative costs).

[b]Refers to the loan amounts at closing that appear in DOE's accounting system. They include the full amount of the loans partially guaranteed through the Financial Institution Partnership Program and do not include capitalized interest.

[c]The Title 17 Innovative Technology Loan Guarantee Program also provided $33 million in credit subsidy costs from fiscal year 2000 through fiscal year 2013 related to other areas such as energy efficiency and electric power transmission projects. These costs are not reflected in this table.

Appendix VII: Information on Other Federal Activities Related to Aspects of U.S. Energy Production and Consumption

This appendix provides more detailed information on other federal activities that were not targeted specifically at fossil, nuclear, or renewable energy production and consumption but may have influenced aspects of U.S. energy production and consumption from 2000 through 2013. It also provides information on federal energy-related research and development (R&D).

Federal Activities Not Targeted at a Specific Energy Source

The studies and reports we reviewed indicated that a number of federal activities may have influenced U.S. energy production and consumption from 2000 through 2013 but were not targeted at a specific energy source. These activities included setting standards and requirements for energy efficiency, selling electricity, providing loans and loan guarantees related to energy efficiency, making outlays for energy consumption and energy efficiency, and forgoing revenues through tax expenditures for electricity transmission and energy efficiency, among other things.

Setting Standards and Requirements

The federal government established or strengthened a number of standards and requirements generally related to energy production and consumption from 2000 through 2013. For example, since the 1970s, the federal government has regulated vehicle fuel economy through corporate average fuel economy (CAFE) standards, which originally required manufacturers to meet a single fleetwide standard for all cars and either a single standard or class standards for light trucks.[1] The Energy Independence and Security Act of 2007 (EISA) instituted several changes to CAFE standards in 2007, such as moving from a single fleet standard to an attribute-based standard.[2] In 2009, the U.S. administration announced a new policy to increase vehicle fuel economy by strengthening CAFE standards and aligning them with the first greenhouse gas emissions standards for vehicles, which would be administered by the Environmental Protection Agency (EPA).[3] The Department of Transportation and EPA established a national program for these two sets of standards by issuing coordinated regulations covering vehicle model years

2012 to 2025 in May 2010 and October 2012.[4] Vehicle manufacturers will have to meet more stringent fuel economy standards, which are projected to be equivalent to over 50 miles per gallon by 2025.[5]

In addition, since the 1970s, the federal government has established minimum efficiency standards requiring that certain products, such as residential appliances, commercial equipment, and lighting products, meet specified energy efficiency standards before they can be sold in the United States.[6] The Energy Policy and Conservation Act of 1975 required the Department of Energy (DOE) to set minimum energy-efficiency standards for manufacturers of specified categories of consumer products such as refrigerators, dishwashers, furnaces, and hot water heaters.[7] The statute was later amended (e.g., by the Energy Policy Act of 2005 [EPAct] and EISA) to include additional categories of consumer products. Manufacturers' compliance with the standards is mandatory. The statute further requires DOE to set and periodically review and revise standards for these product categories to achieve the maximum level of energy efficiency that is technically feasible and economically justified.

The federal standards for vehicle fuel economy and for energy efficiency standards in products provided disincentives for overall energy production and consumption. Specifically, as we concluded in August 2007, the CAFE program reduced oil consumption by cars and light trucks from what it would have otherwise been, and the evidence suggested that increasing CAFE standards would save additional oil in the future.[8] In addition, as we found in March 2013, DOE estimated that, from the inception of the federal minimum efficiency standards program in 1975 through 2005, consumer benefits from these standards amounted to about $64 billion.[9] DOE projected that the standards will save consumers $241 billion by 2030 and $269 billion by 2045.

Directly Providing Goods and Services

The federal government directly provided goods and services generally related to energy production and consumption from 2000 through 2013. For example, the federal government is the largest owner of electricity-generating capacity in the country and owns significant electricity transmission assets. Development of these resources was initially pursued as part of efforts to provide electricity to rural areas, control flooding, and provide irrigation. Five federal utilities—four power marketing administrations and the Tennessee Valley Authority (TVA)— provide electricity and transmission services to customers in their regions.[10]

The power marketing administrations sell power produced primarily at federal hydroelectric dams and projects that are owned and operated by the Department of the Interior's Bureau of Reclamation, the U.S. Army Corps of Engineers, or the International Boundary and Water Commission. TVA markets electricity produced at its own fossil, nuclear, and hydroelectric energy facilities.

In October 2011, we reported on several features of TVA's operations as a federally owned electric utility that provided incentives for energy production and consumption in its power service area—covering about 80,000 square miles in the southeastern United States, including almost all of Tennessee and parts of Mississippi, Kentucky, Alabama, Georgia, North Carolina, and Virginia, with a population of more than nine million people.[11] In fiscal year 2010, TVA sold more than 173 million megawatt-hours of electricity to customers.

To meet this customer demand, TVA generates electricity at 11 coal-fired plants, 11 natural gas-fired plants, 3 nuclear plants, and 29 hydroelectric dams, among other things. Under the TVA Act of 1933, as amended, TVA has not been subject to many of the regulatory oversight requirements that commercial utilities must satisfy.[12]

TVA is also exempt from paying federal and state taxes and can borrow funds for investment in its power system at very competitive interest rates as a result of its triple-A credit rating—which, as we have found, is based partly on its status as a federal entity.[13] Additionally, unlike many utilities, TVA charges rates for its electric power that are not subject to review and approval by state public utility commissions. However, in setting TVA's rates, TVA's Board must comply with the primary objectives of the TVA Act, including the objective that power shall be sold at rates as low as are feasible.

Assuming Risk

Through loan and loan guarantee programs, the federal government assumed risks generally related to energy production and consumption from 2000 through 2013.

Specifically, we identified three federal programs that provided loans and loan guarantees targeting energy efficiency and other activities from fiscal year 2000 through 2013. These programs are described below:

- **Direct and Guaranteed Electric Loan Program.** This U.S. Department of Agriculture (USDA) program, authorized under the Rural Electrification Act of 1936, provides loans and loan guarantees to establish and improve electric service in rural areas, and to assist electric borrowers in implementing demand side management, energy efficiency and conservation programs, and on-grid and off-grid renewable energy systems. These loans and loan guarantees provide financing under favorable terms to eligible nonprofit utility organizations, such as electric co-ops and public utility districts, as well as to for-profit entities. According to USDA, the program supports approximately 700 electric system borrowers in 46 states. This program disbursed about $45 billion in loans from fiscal year 2000 through 2013 with a total cost of about $0.7 billion.[14] To the extent that some of these loans supported energy efficiency and conservation programs, this program provided a disincentive for energy production and consumption. However, because some loans may have supported renewable energy systems, this program also provided incentives for energy production and consumption.

- **Advanced Technology Vehicles Manufacturing loan program.** This DOE program, authorized under EISA, provides loans to support development of advanced technology vehicles and associated components in the United States that would increase the fuel economy of U.S. passenger vehicles. EISA authorized DOE to make $25 billion in loans under this program. As we reported in March 2013, DOE has used the program to make five loans worth $8.4 billion.[15] Two of the loans in this program defaulted, and one has been paid back in full. The current estimated cost of the loans is about $0.3 billion.[16] Because these loans supported improvements in the fuel economy of passenger vehicles, this program provided disincentives for energy production and consumption.
- **Green Retrofit Program for Multifamily Housing.** This Department of Housing and Urban Development program, established by the American Recovery and Reinvestment Act of 2009 (Recovery Act), funds energy and green retrofits to selected affordable multifamily properties through grants and loans. Eligible projects include the installation of efficient heating and cooling systems and appliances, and the upgrade of units to reduce water usage, increase indoor air quality, and provide other various environmental benefits. This program disbursed about $83 million in loans from fiscal year 2000 through 2013 with a total cost of about $66 million.[17]

In addition, we identified other federal loan programs generally related to energy production and consumption, but for which we were unable to identify data on the level of federal support.

For example, the Farm Security and Rural Investment Act of 2002, as amended, established the Rural Energy for America Program within USDA to provide loan guarantees to agricultural producers and small businesses in rural areas to assist with purchasing and installing energy efficiency improvements and renewable energy systems.[18]

Costs associated with loans from this program totaled about $17 million from fiscal year 2000 through 2013; however, USDA did not report costs specifically related to energy efficiency separately from costs related to renewable energy.

Providing Funds

The federal government provided funds generally related to energy production and consumption from 2000 through 2013. Specifically, we identified five federal programs and activities with almost $51 billion in outlays from fiscal year 2000 through 2013. Of these total outlays, about $40 billion were associated with the Department of Health and Human Services' Low Income Home Energy Assistance Program, which provides funds to low-income households to help cover home heating and cooling costs. As shown in figure 27, outlays associated with this program generally increased from fiscal year 2000 through 2010, then decreased through fiscal year 2013. Specifically, program outlays rose from $1.5 billion in fiscal year 2000 to $4.6 billion in fiscal year 2010 before decreasing to $3.5 billion in fiscal year 2013. Because this program made outlays to low-income households to purchase energy, it provided an incentive for energy production and consumption. However, this spending cannot easily be attributed to specific energy sources or fuel types because funds supporting energy purchases in different regions would encourage consumption of different mixes of these fuels, reflecting regional differences in how energy is produced.

Table 8 provides a description of these five federal programs and activities, as well as providing information reported by the Office of Management and Budget (OMB) and provided by DOE on outlays.

We also identified other outlays that generally related to energy production and consumption, but for which we were unable to identify data on the level of federal support. For example, USDA's High Energy Cost Grants program, authorized under the Rural Electrification Act of 1936, provides grants for energy generation, transmission, and distribution facilities serving rural communities with annual average home energy costs that exceed 275 percent of the national average. Applicants may receive grants for on-grid and off-grid renewable energy systems, as well as energy conservation and efficiency projects. This program provided $202 million in outlays from fiscal year 2000 through fiscal year 2013; however, USDA did not report on program outlays specifically related to electricity and energy efficiency separately from outlays related to renewable energy. In general, this program provided incentives for energy production and consumption through its support for energy generation and transmission to rural communities, while providing disincentives for energy consumption through energy conservation and efficiency projects.

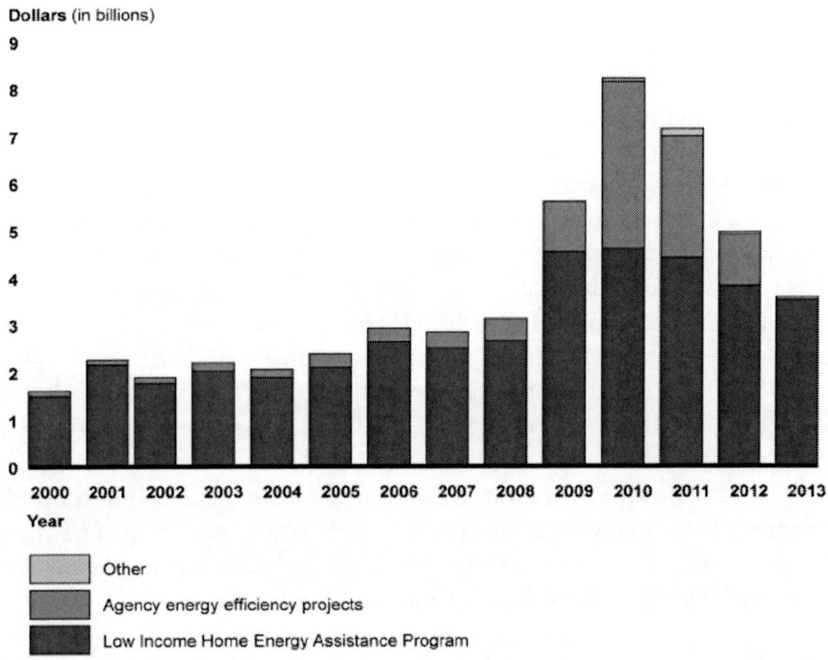

Source: GAO analysis of Office of Management and Budget and DOE data. | GAO-14-836.

Note: Data on agency energy efficiency projects were available only for fiscal years 2000 through 2012. Other includes outlays made by the Department of Housing and Urban Development, Department of the Treasury, and Department of Transportation. Data are not adjusted for inflation.

Figure 27. Federal Outlays Generally Related to Energy Production and Consumption, Fiscal Year 2000 – 2013.

Table 8. Five Federal Programs and Activities with Outlays Generally Related to Energy Production and Consumption, Fiscal Year 2000 – 2013

Name	Description	Outlays
Low Income Home Energy Assistance Program, Department of Health and Human Services	This program was established as a block grant program in 1981 for the purpose of subsidizing heating and cooling costs for low-income households. These grants are provided to states, territories and tribes who are then free to administer the program within the requirements of federal law. In general, federal requirements set eligibility at 150 percent of the federal poverty level or 60 percent of the state's median income. Other federal rules include coordination with DOE's Weatherization Assistance program, annual audits, and outreach activities. Heating and/or cooling assistance funds may be paid directly to eligible households or to retail energy suppliers in the form of cash or vouchers, but in practice the majority of funds are paid directly to the energy providers.	$40.14 billion
Federal agency energy efficiency projectsa	Under legislative mandates and executive orders such as the Energy Policy Act of 2005, Energy Independence and Security Act of 2007, and Executive Order 13423, federal agencies are required to improve the energy efficiency of their buildings. Federal agencies that own or control buildings are required to annually report the energy consumption in these buildings to DOE. DOE collects data from 30 federal agencies, including the Department of Defense, on outlays related to energy efficiency projects, among other things.	$10.34 billionb
Green Retrofit Program for Multifamily Housing, Department of Housing and Urban Development	Established by the Recovery Act, this program funds energy and green retrofits to selected affordable multifamily properties through grants and loans. Eligible projects include the installation of efficient heating and cooling systems and appliances, and the upgrade of units to reduce water usage, increase indoor air quality, and provide other various environmental benefits.	$0.24 billion

Table 8. (Continued)

Name	Description	Outlays
Payment in lieu of tax credit for holders of qualified energy conservation bonds, Department of the Treasury	Issuers of qualified energy conservation bonds can choose to receive a direct payment from the federal government in lieu of tax credit for bondholders.	$0.06 billion
Transit Investments for Greenhouse Gas and Energy Reduction Program, Department of Transportation	This program provides direct funding to public transit agencies for capital investments that assist in reducing the energy consumption or greenhouse gas emissions of public transportation systems. Initiated under the Recovery Act, the program funds capital investments such as energy systems incorporating wind, solar, and geothermal. The program was continued in fiscal years 2010 and 2011. In addition to funding capital investments, the department, through an interagency agreement with DOE's National Renewable Energy Laboratory, is evaluating the overall program, which will include collecting data on projects funded under the program and validating greenhouse gas and energy savings claims made in project proposals.	$0.04 billion

Sources: GAO analysis of Office of Management and Budget and DOE data and documentation. | GAO-14-836.

[a]Federal agencies are also required to improve the energy efficiency of their vehicle fleets, such as by ensuring that most of their vehicle purchases are for alternative fuel vehicles (e.g., electric vehicles and plug-in hybrids). However, we were unable to obtain data on an annual basis related to purchases of energy-efficient vehicles.

[b]DOE data on outlays for agency energy efficiency projects were not available for fiscal year 2013; data are for fiscal years 2000 through 2012.

Forgoing Revenue

The federal government may have influenced general aspects of energy production and consumption from 2000 through 2013 through the tax code. Specifically, we identified 13 tax expenditures that resulted in more than $65 billion in federal revenue losses from fiscal year 2000 through fiscal year 2013. As shown in figure 28, most of these revenue losses ($42 billion or 64 percent) were associated with a tax expenditure excluding employer-paid transportation benefits from taxation, including a number of benefits related to parking, transit passes, and vanpool transportation, among other things. This tax expenditure provided an incentive for the production and consumption of transportation fuels by reducing costs associated with parking. However, it also provided a disincentive for production and consumption of transportation fuels by reducing costs associated with the use of public transportation, which may rely on electricity or other forms of energy.

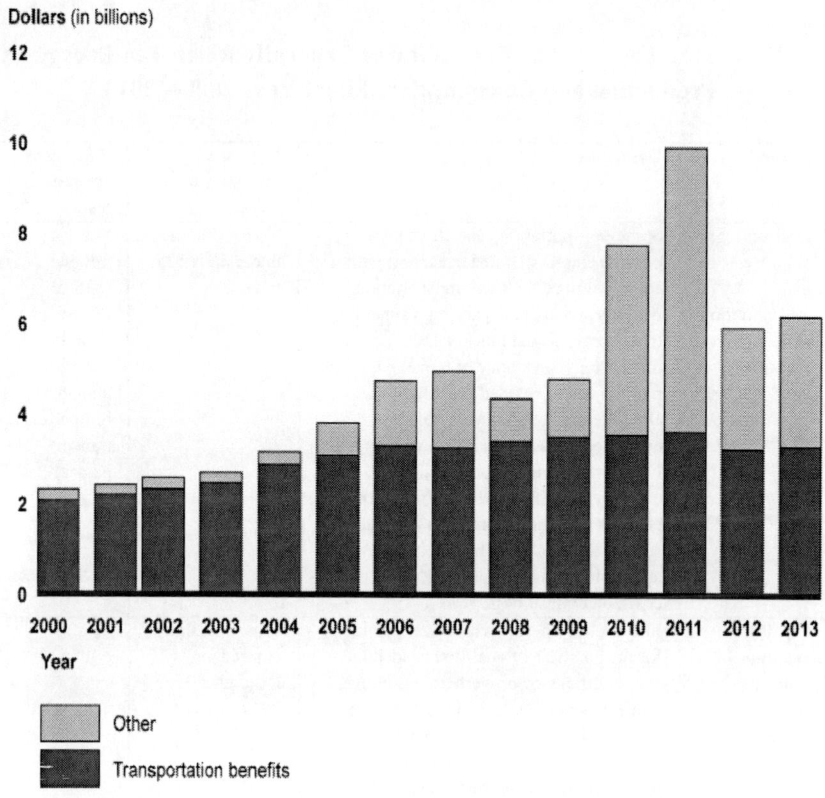

Source: GAO analysis of Treasury and Joint Committee on Taxation data. | GAO-14-836.

Note: Other includes revenue losses associated with 12 tax expenditures we identified as generally related to energy production and consumption. Data are not adjusted for inflation.

Figure 28. Federal Revenue Losses Associated with Tax Expenditures Generally Related to Energy Production and Consumption, Fiscal Year 2000 – 2013.

Table 9 provides descriptions of these 13 federal tax expenditures. It also provides information from the Department of the Treasury (Treasury) on tax expenditures that will or have expired, in full or in part, due to an expiration of legislative authority or some other expiration under the law as of the summer of 2014, as well as on tax expenditures that currently have no expiration. In addition, the table provides information from Treasury on revenue loss estimates (unless otherwise specified).

Table 9. 13 Federal Tax Expenditures Generally Related to Energy Production and Consumption, Fiscal Year 2000 – 2013

Name	Description	Expiration information	Estimated revenue losses
Exclusion of employer-paid transportation benefits (parking, vanpools, and transit passes)	Some transportation benefits employers provide employees are tax exempt within certain limits. Qualified transportation benefits may include parking, vanpool transportation, and transit passes, among other things. The value of transit passes or parking costs provided directly by the employer can be excluded from employees' income, subject to a monthly limit. Transportation provided by employers (as opposed to transportation benefits paid for by employers) is also subject to a qualified tax exclusion. A limit applies to the total of parking, vanpool costs, and transit passes.	No expiration under current law.	$42.13 billiona ($35.82 billion for parking expenses and $6.31 billion for transit passes)
Credit for energy efficiency improvements to existing homes	In 2011, a 10 percent credit was available for the purchase of qualified residential energy efficiency property to encourage homeowners to make their homes more energy efficient. The maximum credit amount was $500, with additional credit limits for specific property. The $500 cap is a lifetime maximum. This credit replaced a 30 percent credit, up to $1,500, that was available during 2009 and 2010. Qualifying energy efficiency improvements include certain improvements to a building's envelope; heating, cooling, and water-heating equipment; and other energy efficiency property.	This provision expired on December 31, 2013.	$10.36 billion

Name	Description	Expiration information	Estimated revenue losses
Credit for residential energy efficient property	A 30 percent tax credit is available for the purchase of residential solar electric property, certain solar water heating property (used for purposes other than heating swimming pools or hot tubs), geothermal heat pumps, small wind energy property, and fuel cell power plants. For fuel cell property, the credit is limited to $500 per half kilowatt of capacity. Otherwise, there is no maximum credit amount. To qualify for the tax credit, eligible property must be installed in the United States in a dwelling used as a residence by the taxpayer. For fuel cell power plants to qualify, they must be installed in connection with the taxpayer's principal residence.	This provision will expire on December 31, 2016.	$3.08 billion
Exclusion of utility conservation subsidies	In general, this provision allows a customer to exclude from their gross income the value of any subsidy provided (directly or indirectly) by a public utility for the purchase or installation of any energy conservation measure. An energy conservation measure is any installation or modification primarily designed to reduce consumption of electricity or natural gas or to improve the management of energy demand with respect to a dwelling unit. To the extent that an energy conservation expenditure qualifies for this exclusion, the taxpayer cannot claim any other tax benefits on the same expenditure.	No expiration under current law.	$2.04 billion
Tax credits for clean-fuel burning vehicles and refueling property	The tax code provides tax credits for a number of vehicles, including alternative fuel vehicles (i.e., vehicles that run on fuels such as compressed natural gas, liquefied natural gas, liquefied petroleum gas, and hydrogen, among other things); fuel cell vehicles; plug-in electric-drive motor vehicles; and two- and three-wheeled plug-in electric vehicles. Tax credits are also available for property used to refuel clean-fuel burning vehicles.	Expiration dates are as follows: • Qualified heavy hybrid vehicles: 12/31/2009. • Fuel cell motor vehicle: 12/31/2014.	$1.77 billion

Table 9. (Continued)

Name	Description	Expiration information	Estimated revenue losses
	A clean fuel refers to any fuel at least 85 percent of the volume of which consists of ethanol or methanol, natural gas, compressed natural gas, liquefied natural gas, liquefied petroleum gas, and hydrogen, among other things.	• Advanced lean burn, qualified hybrid, and qualified alternative fuel motor vehicles: 12/31/2010. • Plug-in conversion credit: 12/31/2011. • 2- or 3-wheeled plug-in electric vehicles, and qualified refueling property: 12/31/2013. The tax code still provides a credit for plug-in electric vehicles, which is based on total manufacturing production and has no expiration date.	
Special rule to implement electric transmission restructuring	The tax code permits taxpayers to elect to recognize any capital gain from the sale of qualifying electricity transmission property to an independent transmission company, pursuant to a Federal Energy Regulatory Commission restructuring policy, evenly over 8 years beginning with the year of the sale. The sale proceeds must be reinvested in other electricity assets within 4 years. Generally, any gain realized from a sale or disposition of a capital asset is recognized in the tax year in which the gain was realized, unless there is a specific exemption or deferral—a taxpayer selling property recognizes any profits for tax purposes in the year of the sale.	This provision expired on December 31, 2013.	$1.68 billion

Name	Description	Expiration information	Estimated revenue losses
	The recognition of gain over 8 years, rather than in the year of sale, is a deferral, rather than a complete forgiveness, of tax liability—it is a delay in the recognition of income, hence in the payment of tax.		
Credit for energy efficient appliances	The tax code provides a tax credit for qualified production (manufacture) of certain energy-efficient dishwashers, clothes washers, and refrigerators to encourage production of appliances that exceed the minimum federal energy-efficiency standards. For example, for dishwashers manufactured in 2011, the per unit credit is as follows: $25 for models that use no more than 307 kilowatt-hours per year and 5.0 gallons per cycle; $50 for models that use no more than 295 kilowatt-hours per year and 4.25 gallons per cycle; and $75 for models that use no more than 280 kilowatt-hours per year and 4 gallons per cycle. Each manufacturer is only eligible for credits for domestic production of energy-efficient units in excess of average production over the past 2 years.	This provision expired on December 31, 2013.	$1.24 billion
Accelerated depreciation recovery periods for specific energy property: electric transmission and smart electric distribution propertyb	A taxpayer is allowed to recover, through annual depreciation deductions, the cost of certain property used in a trade or business or for the production of income. The tax code provides a 15-year recovery period for certain electric transmission property originally placed in service after April 11, 2005. The tax code also provides a 10-year recovery period for qualified smart meter or qualified smart electric grid system (essentially energy monitoring and management devices).	No expiration under current law.	$1.10 billionc
Allowance of deduction for certain energy efficient commercial building property	The tax code provides a formula-based tax deduction for all or part of the cost of energy-efficient commercial building property (i.e., certain major energy-savings improvements made to domestic commercial buildings) placed in service after December 31, 2005, and before January 1, 2014.	This provision expired on December 31, 2013.	$0.78 billion

Table 9. (Continued)

Name	Description	Expiration information	Estimated revenue losses
	The maximum cost of energy-efficient commercial building property that may be deducted in any tax year is limited to the product of $1.80 and the square footage of the building, over deductions claimed for energy efficient commercial building property in any prior tax years. In other words, the deduction is the lesser of: (1) the cost of the energy efficient commercial building property placed in service during the tax year or (2) the product of $1.80 and the square footage of the building, reduced by all deductions claimed with respect to the building in any prior tax years.		
Exclusion of interest on energy facility bonds	Interest income on state and local bonds used to finance the construction of certain private energy facilities is tax exempt. These energy facility bonds are classified as private-activity bonds, rather than as governmental bonds, because a substantial portion of their benefits accrues to individuals or business rather than to the general public. This exemption helps support the financing of projects by local electric utilities because bond purchasers are willing to accept lower before-tax rates of interest than on taxable securities, and these low interest rates enable issuers to provide the services of local energy facilities at lower cost, benefitting end users.	No expiration under current law.	$0.73 billion
Credit for construction of new energy efficient homes	This tax credit provides eligible contractors building energy-efficient new homes with a tax credit of up to $2,000 and eligible manufacturers of manufactured energy-efficient homes with a tax credit of up to $1,000 to encourage the installation of energy-efficient technologies in new homes. Certified energy-efficient new homes qualifying for the tax credit have an annual heating and cooling energy consumption that is at least 50 percent below that of a comparable dwelling unit, among other things.	This provision expired on December 31, 2013.	$0.38 billion

Name	Description	Expiration information	Estimated revenue losses
Five-year carryback period for certain net operating losses of electric utility companies	A net operating loss is generally the amount by which a taxpayer's allowable deductions exceed the taxpayer's gross income. A carryback of such a loss generally results in the refund of federal income tax for the carryback year. A carryover of a loss reduces federal income tax for the carryover year. In general, a loss may be carried back 2 years and carried over 20 years to offset taxable income in such years. The tax code provides an election for certain electric utility companies to extend the carryback period to 5 years for a portion of losses arising in 2003, 2004, and 2005 ("loss years").	Utilities had to make their election during any taxable year ending after December 31, 2005, and before January 1, 2009.	$0.10 billionc
Credit for holding qualified energy conservation bonds	Qualified energy conservation bonds provide an opportunity for tax-exempt entities to issue bonds for which bondholders can receive an income tax credit in lieu of interest payments from the issuers of the bonds. These bonds can be issued to help finance projects that produce or conserve electricity, including capital expenditures incurred for rural development involving the production of electricity from renewable energy resources, as well as qualified facilities. Qualified energy conservation bonds can be used for a broad array of purposes, including expenditures for certain research facilities, grants, and demonstration projects. Similar to the new clean renewable energy bonds, the tax credit rate for qualified energy conservation bonds is 70 percent of the tax credit interest rate. Congress established an initial volume cap of $800 million in bonds that could be issued. The cap was raised to $3.2 billion under the Recovery Act. The Internal Revenue Service allocated the authority to issue the bonds to U.S. states and territories according to a population-based formula, and the states and territories further allocated the bond issuance authority for individual projects.	No expiration under current law.	$0.05 billion

Table 9. (Continued)

Name	Description	Expiration information	Estimated revenue losses
	For bonds issued after March 18, 2010, issuers can elect to receive a direct payment rather than allowing bond purchasers to claim tax credits.		

Sources: GAO analysis of Treasury, Joint Committee on Taxation, and Congressional Research Service information. | GAO-14-836.

Note: Revenue losses reflect Treasury estimates from the President's budget unless otherwise specified. Revenue loss estimates do not incorporate any behavioral responses and thus do not reflect the exact amount of revenue that would be gained if a specific tax expenditure were repealed. In addition, while sufficiently reliable as a gauge of general magnitude, summing individual tax expenditures' revenue loss estimates does not take into account interactions between individual provisions.

[a]The estimate for the exclusion of employer-paid and employer-provided transportation benefits does not include revenue losses associated with employer-provided parking facilities.

[b]The Joint Committee on Taxation generally classifies as tax expenditures cost recovery allowances that are more favorable than those provided under the alternative depreciation system (Internal Revenue Code Section 168(g)), which provides for straight-line recovery over tax lives that are longer than those permitted under the accelerated system. Accelerated depreciation, in effect, reduces the cost of acquiring energy properties by allowing businesses to deduct larger amounts from their taxable income sooner than they would be able to do under straight-line depreciation. Reducing tax liability earlier provides a benefit to the taxpayer because of the time value of money—having a lower tax payment today is worth more to the taxpayer than having the lower payment in the future.

[c]The revenue loss estimate is reported by the Joint Committee on Taxation only.

In addition, we identified other provisions of the tax code that are generally related to energy production and consumption, but for which data on the level of federal support were not readily available. For example, tax-exempt municipal bonds allow publicly-owned utilities to obtain lower interest rates than those available from either private borrowers or the U.S. Treasury. Lower interest rates reduce borrowing costs for such utilities and provide incentives for producing electricity. While tax-exempt municipal bonds are used by energy industries such as electric utilities, the group of eligible borrowers also includes water utilities, telecommunication facilities, waste treatment plants, and other publicly-owned entities. OMB and the Joint

Committee on Taxation did not provide estimates of annual revenue losses related to electric utilities for fiscal year 2000 through 2013 for this provision. Consequently, we cannot report the amount of forgone revenue related to this provision.

Federal R&D Related to Specific Energy Sources

We identified four federal programs that made more than $20 billion in outlays for R&D related to fossil, nuclear, and renewable energy from fiscal year 2000 through fiscal year 2013.[19]

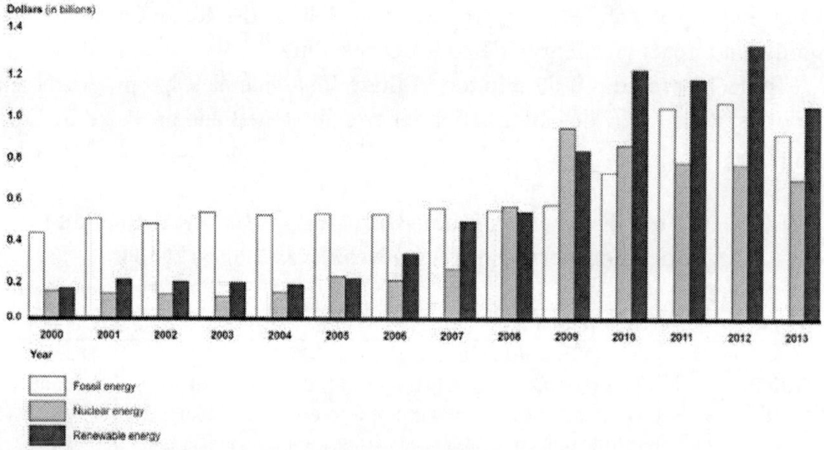

Sources: GAO analysis of Office of Management and Budget. | GAO-14-836.

Note: Some of these outlays may not be related to R&D activities. Fossil energy Includes outlays in DOE's Clean Coal Technology and Fossil Energy Research and Development accounts. Nuclear energy includes outlays in DOE's Nuclear Energy account and outlays in the Energy Supply and Conservation account based on obligated amounts. Renewable energy includes outlays in the Biomass Research and Development account, DOE's Energy Efficiency and Renewable Energy account related to energy supply, and outlays in the Energy Supply and Conservation account based on obligated amounts. Data are not adjusted for inflation.

Figure 29. Federal Outlays for R&D Related to Specific Energy Sources, Fiscal Year 2000 – 2013.

The vast majority of these outlays were made by three DOE program offices—the Office of Fossil Energy, Office of Nuclear Energy, and Office of Energy Efficiency and Renewable Energy. As shown in figure 29, federal outlays related to R&D for fossil, nuclear, and renewable energy generally increased from fiscal year 2000 to 2013, although some variation occurred. Specifically, federal outlays for R&D related to fossil energy increased from about $412 million in fiscal year 2000 to about $1 billion in fiscal year 2012 and $0.9 billion in 2013. Outlays for R&D related to nuclear energy increased over the time period to almost $1 billion in fiscal year 2009, and then decreased through fiscal year 2013. In addition, outlays for R&D related to renewable energy increased to about $1.3 billion in fiscal year 2012 and just over $1 billion in 2013. According to the Energy Information Administration (EIA), new or expanded programs associated with the Recovery Act had a significant impact on energy-related R&D spending.[20]

Table 10 provides a description of these four federal R&D programs and activities; it also provides information reported by OMB and provided by DOE on outlays.

Table 10. Four Federal Programs with Outlays for R&D Related to Specific Energy Sources, Fiscal Year 2000 – 2013

Name	Description	Outlays
Fossil fuel activities, DOE	The Office of Fossil Energy funds R&D and other programs for the reliable, efficient, affordable, and environmentally sound use of fossil fuels, including efforts related to advanced carbon capture and storage for coal-fueled electricity generation. DOE received $3.4 billion from the Recovery Act for fossil fuel R&D.	$8.65 billiona
Renewable energy activities, DOE	The Office of Energy Efficiency and Renewable Energy have funded R&D and other programs to encourage the development and use of renewable energy, including efforts related to biomass systems, geothermal technologies, hydrogen and fuel cell technologies, solar energy technologies, water power, and wind energy.	$7.74 billionb
Nuclear energy activities, DOE	The Office of Nuclear Energy has funded R&D and other programs related to civilian nuclear energy programs, energy technologies, including generation, safety, waste storage and management, and security technologies, to help meet energy security, proliferation resistance, and climate goals.	$5.71 billionc

Name	Description	Outlays
Biomass Research and Development, Department of Agriculture	This program is a joint effort between USDA and DOE to competitively award grants, contracts, and financial assistance to eligible entities to carry out research and development or demonstrations of biofuels and bio-based products and production techniques. Its objectives are to: (1) develop technologies and processes to produce biofuels at prices competitive with fossil fuels, (2) develop high-value bio-based products that can substitute for petroleum-based products, and (3) develop economically and environmentally sustainable domestic sources of renewable biomass for conversion to biofuels, bioenergy, and biobased products.	$0.12 billion

Source: GAO analysis of Office of Management and Budget documentation. | GAO-14-836.

Note: Some of these outlays may not be related to R&D activities. Data are not adjusted for inflation.

[a]Includes outlays in the Clean Coal Technology and Fossil Energy Research and Development accounts. Of this amount, $3.4 billion was appropriated in the Recovery Act in fiscal year 2009. Some of these outlays may not be related to R&D activities.

[b]Includes outlays in DOE's Energy Efficiency and Renewable Energy account related to energy supply, and outlays in the Energy Supply and Conservation account based on obligated amounts.

[c]Includes outlays in DOE's Nuclear Energy account and outlays in the Energy Supply and Conservation account based on obligated amounts.

These outlays provided funding for a variety of energy-specific R&D at federally owned laboratories, as illustrated in the following examples:

- Researchers at DOE's National Energy Technology Laboratory conducted R&D to resolve the environmental, supply, and reliability constraints of producing and using fossil resources. One area of R&D involved combustion science, which provides the basis for a new generation of advanced fossil fuel conversion technologies to meet future demands for efficient, clean, and cost-effective energy production. Combustion science researchers at the laboratory conducted exploratory and applied research in the areas of combustion science technology, and dynamics of engines and other energy conversion devices. This research included modeling, simulation, and laboratory-scale studies of advanced combustion turbines, among other things.

- Researchers at DOE's Idaho National Laboratory conducted R&D on advanced nuclear reactor designs, including the Next Generation Nuclear Power Plant project, which are intended to offer safety and other improvements over the current generation of nuclear power plants. EPAct formally established the Next Generation Nuclear Plant as a DOE project and designated the Idaho National Laboratory as the lead laboratory and construction site for the plant and directs the laboratory to carry out cost-shared R&D, design, and construction activities with industrial partners.
- Researchers at DOE's National Renewable Energy Laboratory (NREL) conducted R&D that led to technological advances for wind and solar energy technologies. For example, laboratory researchers collaborated with wind turbine manufacturers to develop variable-speed turbines to take advantage of lower wind conditions. This innovation allowed a turbine manufacturer to develop and refine its 1.5-megawatt turbines. In addition, laboratory researchers collaborated with solar cell manufacturers to refine their manufacturing techniques before going into full production. This collaboration helped a company to become the world's largest manufacturer of thin-film solar modules.

Federal R&D Generally Related to Energy Production and Consumption

We identified four DOE programs that provided funding for R&D that were not linked to a specific energy source but related to more general aspects of energy production and consumption.[21] These programs made about $47 billion in outlays from fiscal year 2000 through 2013 for R&D related to basic energy sciences, energy efficiency and energy conservation, and electricity grid reliability, among other things.[22] As shown in figure 30, federal outlays for these programs generally increased from fiscal year 2000 to fiscal year 2013, although some variation occurred. Specifically, federal outlays for R&D related to energy efficiency and conservation increased from about $670 million in fiscal year 2000 to over $6 billion in fiscal year 2011 and declining to about $2.0 billion in fiscal year 2013. Outlays for R&D related to basic energy science—funded by DOE's Office of Science—increased from more than $700 million in fiscal year 2000 to about $1.5 billion in fiscal year 2013. In addition, outlays for R&D related to other activities, such as electricity delivery and energy reliability, increased from about $10 million in fiscal year 2003 to about $1.6 billion in fiscal years 2011, before declining to about $1.1 billion in fiscal year 2013. As mentioned above, according to EIA, new or expanded programs associated with the Recovery Act had a significant impact on energy-related R&D spending.

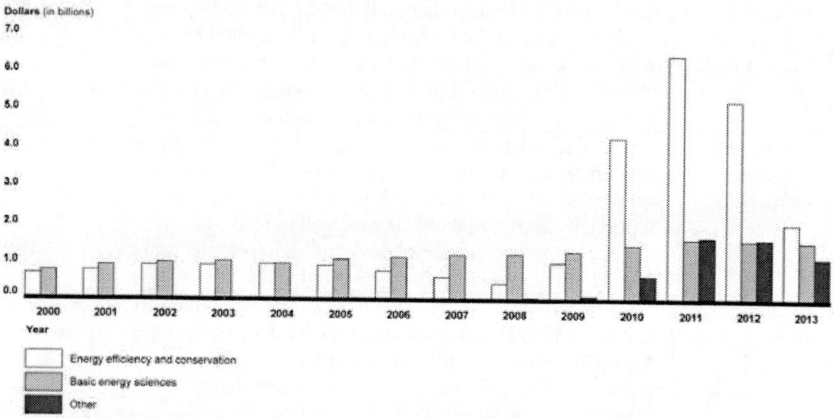

Source: GAO analysis of Office of Management and Budget data. | GAO-14-836.

Note: Other includes outlays associated with DOE's Electricity Delivery and Energy Reliability account and Advanced Research Projects Agency-Energy related accounts. Some of these outlays may not be related to R&D activities. Data are not adjusted for inflation.

Figure 30. Federal Outlays for R&D Generally Related to Energy Production and Consumption, Fiscal Year 2000 – 2013.

Table 11 provides a description of these four federal R&D programs and activities; it also provides information reported by OMB on outlays.

These outlays provided funding for a variety of R&D programs at federally owned laboratories and by nongovernmental entities. For example, researchers at NREL developed a new smart occupancy sensor in 2013 to control lighting and reduce energy costs that could lead to significant energy savings in commercial buildings. In addition, as we found in January 2012, since first receiving an appropriation in the Recovery Act, the Advanced Research Projects Agency-Energy program awarded more than $500 million to universities, public and private companies, and national laboratories to fund 181 projects that attempt to make transformational advances to a variety of energy technologies related to energy efficiency and renewable fuels, among other things.[23]

Table 11. Four Federal Programs with Outlays for R&D Generally Related to Energy Production and Consumption, Fiscal Year 2000 – 2013

Name	Description	Outlays
Energy efficiency and conservation, DOE	The Office of Energy Efficiency and Renewable Energy has funded R&D and other activities for developing and deploying innovative technologies for increasing energy conservation and improving energy efficiency related to vehicle technologies, building technologies, and energy-intensive manufacturing methods, among other things.	$25.52 billiona
Basic energy sciences, DOE	The Office of Science has been the nation's single largest funding source for basic research in energy sciences and supports fundamental research to understand and ultimately control matter and energy at the atomic, molecular, and electronic scales in order to provide the foundations for new energy technologies.	$16.34 billionb
Electricity, DOE	The Office of Electricity Delivery and Energy Reliability has funded R&D and other activities for developing and deploying innovative technologies that improve electric grid reliability, efficiency, flexibility, functionality, and security, among other things.	$4.65 billion
Advanced research projects, DOE	The America Creating Opportunities to Meaningfully Promote Excellence in Technology, Education, and Science Act of 2007 established the Advanced Research Projects Agency-Energy program within DOE to overcome the long-term and high-risk technological barriers in the development of energy technologies.c As specified in statute, the program's goals are to enhance U.S. economic and energy security through the development of certain energy technologies and to ensure that the United States maintains a technological lead in developing and deploying advanced energy technologies.	$0.52 billion

Source: GAO analysis of Office of Management and Budget documentation. | GAO-14-836.

Note: Some of these outlays may not be related to R&D activities. Data are not adjusted for inflation.

[a]Includes outlays in DOE's Energy Efficiency and Renewable Energy, Energy Conservation, and Energy Supply and Conservation accounts under the energy conservation budget subfunction.

[b]Based on outlays from DOE's Science account apportioned using obligated amounts for basic energy sciences.

[c]Pub. L. No. 110-69, § 5012 (2007).

End Notes

[1] See CRS, Energy Tax Policy: Historical Perspectives on and Current Status of Energy Tax Expenditures (Washington, D.C.: May 2, 2011) and Tax-Favored Financing for Renewable Energy Resources and Energy Efficiency (Washington, D.C.: Jan. 10, 2011).

[2] In this report, we use the terms "activity," "support," or "intervention" to refer to a broad range of federal activities that may provide a benefit or preference for a particular activity, product, or outcome. The scope of these terms is intentionally broader than that normally associated with the term "subsidy."

[3] Pub. L. No. 109-58 (Aug. 8, 2005); Pub. L. No. 110-140 (Dec. 19, 2007); and Pub. L. No. 111-5 (Feb. 17, 2009).

[4] EIA is the statistical agency within DOE that collects, analyzes, and disseminates independent information on energy issues. CRS provides information, research, and reference material to assist committees and members of Congress in their legislative and representative functions. CBO assists Congress in carrying out the budget process established by the Congressional Budget Act of 1974. Examples of relevant reports include: EIA, Direct Federal Financial Interventions and Subsidies in Energy in Fiscal Year 2010 (Washington, D.C.: July 2011); CRS, Energy Tax Incentives: Measuring Value Across Different Types of Energy Resources (Washington, D.C.: Sept. 18, 2012); and CBO, Federal Financial Support for the Development and Production of Fuels and Energy Technologies (Washington, D.C.: March 2012).

[5] Crude oil is refined into petroleum products, which include gasoline, diesel, jet fuel, heating oil, and other products.

[6] These categories are based on EIA's methodology for reporting data on renewable energy sources.

[7] Btus are used to measure and compare the energy content of different energy sources. A Btu can be defined as the quantity of heat required to raise the temperature of one pound of water by one degree Fahrenheit. One quadrillion is equal to one thousand trillion. In physical terms, one quadrillion Btus is equivalent to the energy contained in 172 million barrels of crude oil, which is about how much petroleum the United States consumes in 9 days.

[8] Imported energy refers to net imports, or the amount of energy imported minus the amount exported. EIA updates historical energy data periodically. We used EIA data from the spring of 2014; in some cases, the data we collected for this report may not match revised EIA data.

[9] According to EIA officials, while it is possible to capture some of this waste energy, there are theoretical limits to this capture, and it is capital intensive, which adds to an already capital intensive generation facility. Given when these plants were built, they would not be considered "inefficient" in an economic sense.

[10] In addition to conversion losses, other losses include power plant use of electricity, as well as transmission and distribution of electricity from power plants to end-use consumers (also called "line losses"). See EIA, Monthly Energy Review (Washington, D.C.: November 2013).

[11] Components of gross domestic product include personal consumption expenditures, gross private domestic investment, net exports of goods and services, and government consumption expenditures and gross investment.

[12] For these units, dollars refer to a dollar value adjusted to remove the effects of inflation by dividing the nominal value (also called the current dollar value) by the appropriate price index (2009 in this case).

[13] According to USDA officials, changes in consumer behavior and demographic trends may have also affected energy consumption.

[14] International Energy Agency, The Impact of the Financial and Economic Crisis on Global Energy Investment: IEA Background Paper for the G8 Energy Ministers' Meeting: 24-25 May 2009.

[15] In this context, the term "regulation" refers to rules governing private activities rather than those governing internal agency operations.

[16] Pub. L. No. 83-703, 68 Stat. 919, 42 U.S.C. § 2011 et seq.

[17] Clean Air Act Amendments of 1970, Pub. L. No. 91-604, 84 Stat.1676 (1970) (codified as amended at 42 U.S.C. §§ 7401-7671q (2011) (commonly referred to as the Clean Air Act).

[18] See appendix II for a more information on federal sources of data for energy-related outlays.

[19] Other types of taxes include social insurance taxes (employee and employer contributions for Social Security, Medicare, and unemployment compensation); and estate and gift taxes. Excise taxes are generally collected by businesses, which remit payments to the government on a quarterly basis, on selected items such as gasoline, alcoholic beverages, and airline tickets.

[20] The outer continental shelf consists of submerged federal lands, generally extending seaward between 3 and 200 nautical miles off the coastline. EIA estimated that 28 percent of all fossil energy produced in the United States in 2012 was obtained on federal lands (including submerged lands in the outer continental shelf).

[21] The concept of tax expenditures extends beyond the income tax. Tax expenditures also exist for other types of taxes, such as excise and payroll taxes.

[22] See appendix II for a more information on federal sources of data for energy-related tax expenditures.

[23] Pub. L. No. 93-344, § 3, 88 Stat. 297, 299 (July 12, 1974), codified at 2 U.S.C. § 622(3).

[24] The ultimate cost of tax expenditures in forgone revenues is hard to quantify because it is up to individuals and corporations to claim benefits from the provisions of the tax code that apply to them. As a result, the level of forgone federal revenues from tax expenditures can only be estimated. In addition, revenue loss estimates do not represent the amount of revenue that would be gained if a particular tax provision were repealed, since repeal of the tax provision would probably change taxpayer behavior in some way that would affect revenue.

[25] Pub. L. No. 104-58, § 304, 109 Stat. 563, 565 (Nov. 28, 1995).

[26] For purposes of this report, we included activities related to technology demonstration as part of R&D and excluded activities related to commercialization and deployment. Commercialization includes efforts to transition technologies to commercial applications by bridging the gap between research and demonstration activities and venture capital funding and marketing activities. Deployment includes efforts that facilitate or achieve widespread use of technologies in the commercial market.

[27] For more information, see GAO, National Laboratories: DOE Needs to Improve Oversight of Work Performed for Non-DOE Entities, GAO-14-78 (Washington, D.C.: Oct. 25, 2013).

[28] GAO, Renewable Energy: Federal Agencies Implement Hundreds of Initiatives, GAO-12-260 (Washington, D.C.: Feb. 27, 2012).

[29] GAO, Batteries and Energy Storage: Federal Initiatives Supported Similar Technologies and Goals but Had Key Differences, GAO-12-842 (Washington, D.C.: Aug. 30, 2012).

[30] Natural gas and crude oil are found in a variety of geological formations. Conventional natural gas and crude oil are found in deep, porous rock or reservoirs and can flow under natural pressure to the surface after drilling. In contrast, the low permeability of some formations, including shale, means that natural gas and crude oil trapped in the formation cannot move

easily within the rock. Tight formations refer to low permeability formations that include shale as well as sandstones and carbonates.

[31] Liquefied natural gas is natural gas that has been cooled to -260 degrees Fahrenheit. This cooling shrinks the volume of the natural gas by over 600 times, allowing greater quantities of natural gas to be shipped in specially built vessels.

End Notes for Appendix I

[1] Tax expenditures are reductions in a taxpayer's tax liability as a result of special exemptions and exclusions from taxations, deductions, credits, deferrals of tax liability, or preferential tax rates.

[2] Specifically, we used the most recent tax expenditure estimates developed by Treasury and reported by OMB in the Federal Budget for fiscal years 2002 to 2015 and JCT estimates reported in its Estimates of Federal Tax Expenditures reports for fiscal years 2000-2012. For fiscal year 2013, we used estimates from the 2012 JCT report, which reflect the provisions in federal tax law enacted through January 2, 2013.

[3] Treasury and JCT report estimates only for income tax expenditures and do not report estimates for tax provisions that result in forgone excise tax only. Treasury and JCT use different (1) income tax baselines, (2) de minimis amounts (which is the minimum revenue loss threshold for Treasury and JCT to report a tax expenditure), and (3) economic and technical assumptions. For more information on how Treasury and JCT estimate revenue loss, see appendix III in GAO, Government Performance and Accountability: Tax Expenditures Represent a Substantial Federal Commitment and Need to Be Reexamined, GAO-05-690 (Washington, D.C.: Sept. 23, 2005).

End Notes for Appendix II

[1] GAO, Renewable Energy: Federal Agencies Implement Hundreds of Initiatives, GAO-12-260 (Washington, D.C.: Feb. 27, 2012).

[2] In this report, we use the term "revenue loss" to mean forgone revenue, or revenue that would have accrued to the federal government had the tax expenditure or other incentive not been provided for in the law.

End Notes for Appendix III

[1] In 2013, the United States imported about 2.8 trillion cubic feet of natural gas from Canada and exported about 0.9 trillion cubic feet of natural gas to Canada and 0.7 trillion cubic feet to Mexico.

[2] GAO, Energy-Water Nexus: Information on the Quantity, Quality, and Management of Water Produced during Oil and Gas Production, GAO-12-156 (Washington, D.C.: Jan. 9, 2012). See also GAO, Oil and Gas: Information on Shale Resources, Development, and Environmental and Public Health Risks, GAO-12-732 (Washington, D.C.: Sept. 5, 2012).

[3] CRS, Natural Gas in the U.S. Economy: Opportunities for Growth (Washington, D.C.: Nov. 6, 2012).

[4] CRS, Natural Gas in the U.S. Economy: Opportunities for Growth.

[5] A spot market for natural gas is a market in which natural gas is bought and sold for immediate or very near-term delivery, usually for a period of 30 days or less. A spot market is more likely to develop at a location with numerous pipeline interconnections, thus allowing for a large number of buyers and sellers.

[6] CRS, Natural Gas in the U.S. Economy: Opportunities for Growth.

[7] Most major importers of liquefied natural gas are non-free-trade-agreement countries. DOE is responsible for approving the export of liquefied natural gas as a commodity. Before approving applications to export liquefied natural gas to countries without free trade agreements, DOE must determine that the exports are consistent with the public interest.

[8] The Federal Energy Regulatory Commission is responsible for authorizing the construction and operation of facilities that can liquefy and export natural gas, and as of July 2014, had approved three liquefied natural gas export terminals. In the latter half of the 2000s, companies constructed numerous liquefied natural gas import facilities in the United States, and many of the proposed export facilities will be constructed at the site of these import facilities. GAO recently reported on DOE's process for reviewing export applications and the Federal Energy Regulatory Commission's process for reviewing export facility applications. See GAO, Natural Gas: Federal Approval Process for Liquefied Natural Gas Exports, GAO-14-762 (Washington, D.C.: Sept. 26, 2014).

[9] EIA, Annual Energy Outlook 2013 (Washington, D.C.: April 2013).

[10] A refinery's capacity refers to the maximum amount of crude oil designed to flow into the distillation unit of a refinery, also known as the crude unit.

[11] We reported in 2014 on several other factors that may have also contributed to increasing petroleum product exports, including increases in U.S. production of crude oil and natural gas that have lowered costs and provided a competitive advantage to U.S. refiners. See Petroleum Refining: Industry's Outlook Depends on Market Changes and Key Environmental Regulations, GAO-14-249 (Washington, D.C.: Mar. 14, 2014).

[12] GAO-14-249. Oil and natural gas are often found together in the same reservoir. The natural gas produced from oil wells is generally classified as "associated-dissolved," meaning that it is associated with or dissolved in crude oil. In areas where the primary purpose of drilling is to produce oil, operators may flare associated natural gas because no local market exists for the gas and transporting to a market may not be economically feasible. In September 2012, we found that flaring poses a risk to air quality because it emits carbon dioxide, a greenhouse gas linked to climate change. For more information, see GAO-12-732.

[13] West Texas Intermediate refers to a crude oil stream produced in Texas and southern Oklahoma that serves as a reference for pricing a number of other crude streams. Brent refers to a blended crude stream produced in the North Sea region that serves as a reference for pricing a number of other crude streams. A crude oil stream refers to crude oil produced in a particular field or a collection of crude oils with similar qualities from fields in close proximity.

[14] EIA, Short-Term Energy Outlook Supplement: Brent Crude Oil Spot Price Forecast (Washington, D.C.: July 10, 2012).

[15] Crude oil pipelines connect several large refining centers to crude oil sources, and petroleum product pipelines connect these refineries to population centers all over the country. According to EIA, because of these pipeline constraints, oil producers in the midcontinental United States had to move additional volumes by truck or rail. We recently found that

expansions in U.S. infrastructure for oil and gas transportation have not kept pace with increased domestic oil and gas production. See GAO, Oil and Gas Transportation: Department of Transportation Is Taking Actions to Address Rail Safety, but Additional Actions Are Needed to Improve Pipeline Safety, GAO-14-667 (Washington, D.C.: Aug. 21, 2014).

[16] EIA, Today in Energy: U.S. Crude Oil Production Growth Contributes to Global Oil Price Stability in 2013 (Washington, D.C.: Jan. 9, 2014).

[17] EIA, Energy in Brief: What Is the Role of Coal in the United States? (Washington, D.C.: Aug. 16, 2013).

[18] A short ton is 2,000 pounds.

[19] A kilowatt-hour is a unit of work or energy equal to 1 kilowatt (or 1,000 watts) of power expended for 1 hour.

[20] EIA, Energy in Brief: What Is the Role of Coal in the United States?

[21] See GAO, Electricity: Significant Changes Are Expected in Coal-Fueled Generation, but Coal Is Likely to Remain a Key Fuel Source, GAO-13-72 (Washington, D.C.: Oct. 29, 2012).

[22] For more information, see GAO, EPA Regulations and Electricity: Update on Agencies' Monitoring Efforts and Coal-Fueled Generating Unit Retirements, GAO-14-672 (Washington, D.C.: Aug. 15, 2014).

[23] EIA, Today in Energy: Subbituminous and Bituminous Coal Dominate U.S. Coal Production (Washington, D.C.: Aug. 16, 2011). Other types include lignite and anthracite, both of which comprise a small percentage of total U.S. production.

[24] See EIA, Annual Energy Outlook: 2010 (Washington, D.C.: May 11, 2010); and Today in Energy: Cost of Transporting Coal to Power Plants Rose Almost 50% in Decade (Washington, D.C.: Nov. 19, 2012).

[25] Clean Air Act Amendments of 1970, Pub. L. No. 91-604, 84 Stat.1676 (1970) (codified as amended at 42 U.S.C. §§ 7401-7671q (2011) (commonly referred to as the Clean Air Act).

[26] EPA has set national ambient air quality standards for six pollutants, termed "criteria" pollutants: carbon monoxide, lead, nitrogen oxides, ozone, particulate matter, and sulfur oxides.

[27] GAO, Air Pollution: EPA Needs Better Information on New Source Review Permits, GAO-12-590 (Washington, D.C.: June 22, 2012).

[28] Pub. L. No. 101-380, 104 Stat. 484 (1990).

[29] The federal government maintains other trust funds to hold funds for a specific purpose that may be related to the production and consumption of fossil energy. For example, the Leaking Underground Storage Tank Trust Fund provides funding to clean up leaks of petroleum and other substances from underground storage tanks. (26 U.S.C. § 9508).

[30] The barrel tax expired in December 1994 and was reinstituted at 5 cents per barrel in April 2006 as mandated by the Energy Policy Act of 2005 (Pub. L. No. 109-58, §1361, 119 Stat. 594, 1058 (2005)). The Energy Improvement and Extension Act of 2008 (Pub. L. No. 110-343, § 405, 122 Stat. 3765, 3860 codified as amended at 26 U.S.C. § 4611 (c)(2)(B)) increased the tax to 8 cents per barrel (and 9 cents per barrel in 2017) and provided that the fund's barrel tax shall expire after December 31, 2017.

[31] Energy Policy and Conservation Act, Pub. L. No. 94-163, §161, 89 Stat. 888-889 (1975), codified as amended at 42 U.S.C. § 6241. The statute provides for a drawdown of the reserve upon a finding by the President that there is a "severe energy supply interruption" as well as an event that is, or is likely to become, an energy supply shortage "of significant scope or duration." 42 U.S.C. § 6241(d), (h).

[32] The Strategic Petroleum Reserve also received crude oil through the royalty-in-kind transfer program established by the Department of the Interior. The program applied to crude oil that was produced from federal leases in the Gulf of Mexico and was transferred to the U.S. government in lieu of cash royalty payments. According to Department of the Interior data, the value of the forgone royalty receipts totaled about $6.5 billion from fiscal years 2000 through 2009 when the program was terminated.

[33] This tax generated about $2.8 billion for the fund from 2000 through 2012, the most recent data reported by the Internal Revenue Service. The tax is scheduled to expire to expire in 2017, putting the federal government's longer-term ability to provide financial support in response to oil spills at risk, as we found in October 2011. See GAO, Deepwater Horizon Oil Spill: Actions Needed to Reduce Evolving but Uncertain Federal Financial Risks, GAO-12-86 (Washington, D.C.: Oct. 24, 2011).

[34] GAO, Strategic Petroleum Reserve: Available Oil Can Provide Significant Benefits, but Many Factors Should Influence Future Decisions about Fill, Use, and Expansion, GAO-06-872 (Washington, D.C.: Aug. 24, 2006).

[35] Data for excise taxes collected in 2012 are the most recent data available from the Internal Revenue Service. These data comprise revenues from taxes on a variety of motor fuels, including gasoline, diesel, kerosene, aviation-grade kerosene, liquefied natural gas, liquefied fuel derived from coal, liquefied hydrogen, compressed natural gas, liquid petroleum gas, and others.

[36] Of the 18.4 cent per gallon tax, 0.1 cents is dedicated to the Leaking Underground Storage Tank Trust Fund; the remainder is dedicated to the Highway Trust Fund.

[37] Revenues consist of royalties, rents, bonuses, and other revenues that mineral royalty payers report to the Department of the Interior. According to Interior documentation, comparable data were not available for 2000 through 2002; as a result, we are unable to report data on royalty and other payments for those years.

[38] The table also provides information from Treasury on tax provisions that will or have expired, in full or in part, due to an expiration of legislative authority or some other expiration under the law as of the fall of 2014.

[39] These estimated revenue losses include those from royalty relief for lease sales from 1996 through 2000 as mandated by the Outer Continental Shelf Deep Water Royalty Relief Act and reflect forgone revenues in the calendar year in which the sale of product took place.

[40] Compared with other countries within the Organization for Economic Cooperation and Development, the United States has one of the lowest excise tax rates for motor fuels.

[41] GAO, Oil and Gas Royalties: The Federal System for Collecting Oil and Gas Revenues Needs Comprehensive Reassessment, GAO-08-691 (Washington, D.C.: Sept. 3, 2008).

End Notes for Appendix IV

[1] A megawatt-hour is a unit of work or energy equal to 1 megawatt (or one million watts of power) expended for 1 hour. One megawatt provides enough electricity to power about 750 homes.

[2] NRG, News Release: NRG Energy, Inc. Provides Greater Clarity on the South Texas Nuclear Development Project (STP 3&4), Apr. 19, 2011.

[3] The task force concluded that a sequence of events similar to that at Fukushima was unlikely to occur in the United States and that the continued operation of the nation's commercial nuclear reactors did not pose an imminent risk to public health and safety.

[4] Although DOE was required to take custody of spent nuclear fuel from commercial nuclear reactor operators for disposal at the repository beginning in January 1998, DOE was unable to begin receiving waste by that time because of a series of delays due to, among other reasons, state and local opposition to the construction of a permanent nuclear waste repository in Nevada and technical complexities.

[5] GAO, Spent Nuclear Fuel: Accumulating Quantities at Commercial Reactors Present Storage and Other Challenges, GAO-12-797 (Washington, D.C.: Aug. 15, 2012).

[6] NRC's Commission has five members who are responsible for formulating policies and regulations governing safety and security of nuclear reactors and materials, issuing orders to licensees, and adjudicating legal matters brought before it.

[7] In August 2013, the U.S. Court of Appeals for the District of Columbia Circuit directed NRC to resume the licensing process for DOE's Yucca Mountain repository application. In re Aiken County, 725 F.3d 255 (D.C. Cir. 2013), reh'g en banc denied (Oct.28, 2013). In response to the court's ruling, NRC directed its staff to complete and issue the safety evaluation report associated with DOE's license application.

[8] Pub. L. No. 85-256, § 4, 71 Stat. 576, 576-79 (codified as amended at 42 U.S.C. § 2210).

[9] The insurance coverage has two layers: The owner of a nuclear plant is required to purchase primary insurance covering liability up to $375 million. In the event of an accident, liability for damages assessed at between $375 million and $13 billion would then be shared among the owners of all U.S. nuclear plants, who would pay a "retroactive premium."

[10] Pub. L. No. 109-58, § 1703, 119 Stat. 660, 1120 (2005) codified as amended at 42 U.S.C. §16513.

[11] DOE issued about $6.2 billion of these loan guarantees in February 2014. For more information, see GAO, DOE Loan Programs: DOE Should Fully Develop Its Loan Monitoring Function and Evaluate Its Effectiveness, GAO-14-367 (Washington, D.C.: May 1, 2014).

[12] CBO, Nuclear Power's Role in Generating Electricity (Washington, D.C.; May 2008).

[13] These estimates are from the Joint Committee on Taxation. We identified an additional tax expenditure related to nuclear energy—the advanced nuclear power production credit. However, there were no revenue losses associated with this tax expenditure from fiscal year 2000 through 2013, according to the Office of Management and Budget.

[14] The special decommissioning funds were first established by the Deficit Reduction Act of 1984, but the funds' investment earnings were initially subject to tax at the highest corporate tax rate (46 percent at the time). Subsequent amendments to the provision have reduced the tax rate to 20 percent.

End Notes for Appendix V

[1] According to EIA, production equals consumption for all renewable energy sources except biofuels.

[2] According to EIA, over 99 percent of ethanol used in the United States is consumed as E10, a blend of 10 percent ethanol and 90 percent gasoline by volume. A small amount of ethanol

is sold in high blends, such as E15 (a blend of 15 percent ethanol and 85 percent gasoline by volume) and E85 (a blend of 51 to 83 percent ethanol by volume and gasoline).

[3] In recent years, the United States also increased imports of ethanol from Brazil to meet renewable fuel requirements for advanced biofuels. Brazilian ethanol is made from sugarcane and qualifies as an advanced biofuel, while domestic ethanol produced from corn does not.

[4] The 45-cents-per-gallon tax credit is also referred to as the volumetric ethanol excise tax credit. The federal government also provided tax credits (and a related excise tax credit) for biodiesel. See appendix VI for more information on the alcohol fuel credits and biodiesel credits.

[5] The alcohol fuel credits generally expired in 2011, but some taxpayers were still able to claim the credit in 2012 and 2013 due in part to the timing and method of taxpayer filing and Internal Revenue Service processing, according to Treasury officials.

[6] Pub. L. No. 109-58, § 1501, 119 Stat. 660, 1067 (2005). States have also played a role in influencing the increased production and consumption of ethanol. According to the American Coalition for Ethanol, 10 states have enacted standards that require the use of ethanol-blended fuel, while 22 states have established some type of incentive for ethanol producers.

[7] Pub. L. No. 110-140, § 201, 121 Stat. 1492, 1519 (2007).

[8] GAO, Biofuels: Potential Effects and Challenges of Required Increases in Production and Use, GAO-09-446 (Washington, D.C.: Aug. 25, 2009).

[9] See also GAO, Opportunities to Reduce Potential Duplication in Government Programs, Save Tax Dollars, and Enhance Revenue, GAO-11-318SP (Washington, D.C.: Mar. 1, 2011).

[10] The cellulosic biofuel credit—part of the alcohol fuel credits—was extended through December 31, 2013.

[11] Section 701 of the Energy Policy Act of 2005, (42 U.S.C. § 6374(a)(3)(E)), provides for a waiver to this requirement if in a particular geographic area alternative fuel is not reasonably available or the cost of the alternative fuel is unreasonably more expensive compared to gasoline.

[12] Pub. L. No. 110-140, § 142, 121 Stat. 1492, 1518 (2007).

[13] The excise tax is on "gasohol," which is a blend of finished motor gasoline containing alcohol (generally ethanol but sometimes methanol) at a concentration between 5.7 percent and 10 percent by volume.

[14] Lihong McPhail and Xiaodong Du, “Ethanol Strengthens the Link between Agriculture and Energy Markets," in USDA Economic Research Service, Amber Waves, June 5, 2012.

[15] DOE, 2012 Wind Technologies Market Report (Washington, D.C.: August 2013).

[16] GTM Research and Solar Energy Industries Association, U.S. Solar Market Insight Report: 2012 Year in Review: Executive Summary (2013).

[17] EIA, Annual Energy Outlook 2014: State Renewable Energy Requirements and Goals: Update through 2013 (Washington, D.C.: Apr. 30, 2014).

[18] CRS, U.S. Renewable Electricity: How Does the Production Tax Credit (PTC) Impact Wind Markets? (Washington, D.C.: Oct. 10, 2012).

[19] The production tax credit expired on July 31, 1999; Dec. 31, 2001; and Dec. 31, 2003. It was also scheduled to expire on Dec. 31, 2012; however, the American Taxpayer Relief Act of 2012 extended the credit for wind projects that began construction before Jan. 1, 2014.

[20] GAO, Wind Energy: Additional Actions Could Help Ensure Effective Use of Federal Financial Support, GAO-13-136 (Washington, D.C.: Mar. 11, 2013).

[21] The investment tax credit also provides a tax credit for geothermal systems, fuel cells, microturbines, and combined heat and power.

[22] Treasury rounds all yearly estimates to the nearest $10 million and excludes tax expenditures with estimates that round to zero in each of the 7 years that it reports tax expenditure estimates.

[23] EPAct also required installation of 20,000 solar energy systems in federal buildings by 2010.

[24] These estimated costs are credit subsidy costs, which represent the government's estimated net long-term cost of extending or guaranteeing credit, in present value terms, over the entire period the loans are outstanding (not including administrative costs).

[25] DOE, SunShot Vision Study (Washington, D.C.: February 2012).

End Notes for Appendix VII

[1] The Energy Policy and Conservation Act, Pub. L. No. 94-163, 89 Stat. 871 (Dec. 22, 1975), established CAFE standards beginning in 1978 (for passenger cars).

[2] Pub. L. No. 110-140, 121 Stat. 1492 (Dec. 19, 2007). An attribute-based standard is based on a vehicle's "footprint," or the size of its wheelbase multiplied by its average track width.

[3] When vehicles burn gasoline and other fossil fuels, they emit greenhouse gases such as carbon dioxide, methane, nitrous oxide, and hydrofluorcarbons, according to EPA. EPA has established standards for the amount of greenhouse gases that can be emitted from vehicles.

[4] 75 Fed. Reg. 25324 (May 7, 2010); and 77 Fed. Reg. 62624 (Oct. 15, 2012).

[5] Specifically, EPA is establishing standards that are projected to require, on an average industry fleetwide basis, 163 grams/mile of carbon dioxide in model year 2025 vehicles, which is equivalent to 54.5 miles per gallon if this level were achieved solely through improvements in fuel efficiency.

[6] State governments also played an important role in encouraging efforts to use energy more efficiently. According to the American Council for an Energy-Efficient Economy, as of April 2014, 25 states had fully funded policies in place that establish specific energy savings targets that utilities or nonutility program administrators must meet through customer energy efficiency programs.

[7] Pub. L. No. 94-163, 89 Stat. 871, 917-932 (Dec. 22, 1975), (codified as amended at 42 U.S.C. §§ 6291-6309).

[8] GAO, Vehicle Fuel Economy: Reforming Fuel Economy Standards Could Help Reduce Oil Consumption by Cars and Light Trucks, and Other Options Could Complement These Standards, GAO-07-921 (Washington, D.C.: Aug. 2, 2007).

[9] GAO, Energy Efficiency: Better Coordination among Federal Programs Needed to Allocate Testing Resources, GAO-13-135 (Washington, D.C.: Mar. 28, 2013).

[10] The power marketing administrations are the Bonneville Power Administration, Southeastern Power Administration, Southwestern Power Administration, and Western Area Power Administration. TVA is a wholly-owned government corporation.

[11] GAO, Tennessee Valley Authority: Full Consideration of Energy Efficiency and Better Capital Expenditures Planning Are Needed, GAO-12-107 (Washington, D.C.: Oct. 31, 2011).

[12] As an electric utility, TVA is subject to certain aspects of Federal Energy Regulatory Commission jurisdiction, as well as to other regulatory bodies and agencies including the Nuclear Regulatory Commission and EPA.

[13] While TVA does not pay federal and state taxes, it does make payments-in-lieu-of-taxes to state and local governments.

[14] These costs are current estimates of the credit subsidy costs of disbursed amounts. Credit subsidy costs represent the government's estimated net long-term cost of extending or guaranteeing credit, in present value terms, over the entire period the loans are outstanding (not including administrative costs).

[15] GAO, Department of Energy: Status of Loan Programs, GAO-13-331R (Washington, D.C.: Mar. 15, 2013).

[16] These costs are current estimates of the credit subsidy costs of disbursed amounts. [17]These costs are current estimates of the credit subsidy costs of disbursed amounts.

[18] This program was formerly called the Renewable Energy Systems and Energy Efficiency Improvements Program.

[19] Some outlays associated with these programs may not be related to R&D. Other federal agencies may have provided outlays related to R&D for specific energy sources during this time period; however, data associated with these outlays were not readily available.

[20] EIA, Direct Federal Financial Interventions and Subsidies in Energy in Fiscal Year 2010 (Washington, D.C.: July 2011).

[21] Other federal agencies may have provided outlays for R&D generally related to energy production and consumption during this time period; however, data associated with these outlays were not readily available.

[22] Some outlays associated with these programs may not be related to R&D.

[23] GAO, Department of Energy: Advanced Research Projects Agency-Energy Could Benefit from Information on Applicants' Prior Funding, GAO-12-112 (Washington, D.C.: Jan. 13, 2012).

In: U.S. Energy Production and Consumption ISBN: 978-1-63482-182-7
Editor: Kendra Fletcher

Chapter 2

ENERGY POLICY: 113TH CONGRESS ISSUES*

Brent D. Yacobucci

SUMMARY

Energy policy in the United States has focused on three major goals: assuring a secure supply of energy, keeping energy costs low, and protecting the environment. In pursuit of those goals, government programs have been developed to improve the efficiency with which energy is utilized, to promote the domestic production of conventional energy sources, and to develop new energy sources, particularly renewable sources.

Implementing these programs has been controversial because of varying importance given to different aspects of energy policy. For some, dependence on imports of foreign oil, particularly from the Persian Gulf, is the primary concern; for others, the indiscriminate use of fossil fuels, whatever their origin, is most important. The contribution of burning fossil fuels to global climate change is particularly controversial. Another dichotomy is between those who see government intervention as a positive force and those who view it as a necessary evil at best.

In the 113th Congress, energy legislation has generally been focused on specific topics, as opposed to broad legislation like that enacted in

* This is an edited, reformatted and augmented version of a Congressional Research Service publication R42756, prepared for Members and Committees of Congress, dated September 18, 2014.

1992, 2005, and 2007. Congress has enacted statutes on generally less controversial topics including hydroelectric licensing and permitting and pipeline safety documentation (P.L. 113-30). The 113th Congress also enacted the Water Resources Reform and Development Act of 2014 (WRRDA, P.L. 113-121), omnibus reauthorization of water resource activities. A number of issues, some of which drew attention in previous Congresses, have been taken up in proposed legislation.

The Energy Savings and Industrial Competitiveness Act of 2013, S. 2262and H.R. 1616, would promote energy efficiency in buildings and industry by encouraging adoption of uniform building codes and authorize a grant program for state energy efficiency programs. Several versions of the bill have seen committee and/or floor action in the Senate, but S. 2262 failed a cloture vote on May 12, 2014. On March 5 the House passed H.R. 2126, which contains some of the provisions in S. 2262, by a vote of 375-36.

H.R. 3, the Northern Route Approval Act, would declare that a presidential permit would not be required for construction of the Keystone XL pipeline from Canada to the Gulf of Mexico. The bill passed the House on May 22, 2013, by a vote of 241-175. The issue of approving the Keystone XL pipeline is deeply involved in the question of global climate change, since opponents argue that Canadian oil sands production contributes excessive amounts of greenhouse gas emissions.

A bill promoting increased offshore oil and gas exploration (H.R. 2231) passed the House on June 28, 2013. In November 2013 the House passed three energy related bills. H.R. 2728 would prohibit the Bureau of Land Management from enforcing rules to regulate hydraulic fracturing in any state that has its own regulations on the activity. H.R. 1900 would require the Federal Energy Regulatory Commission to respond to permit applications for new natural gas pipelines within one year. H.R. 1965 would be aimed at increasing on-shore leasing of oil and gas resources on federal land. On March 6, 2014, the House passed H.R. 3826, a bill to block the Environmental Protection Agency from limiting power plant carbon emissions.

On September 16, 2014, the House Rules Committee met to consider a floor rule for H.R. 2, which contains provisions from several bills, including H.R. 3, H.R. 1900, H.R. 1965, H.R. 2231, and H.R. 3826. Floor debate on the bill has was scheduled for September 18.

INTRODUCTION

U.S. energy policy since the Arab oil embargo in the 1970s has been aimed at a long-term goal with three major dimensions: to assure a secure supply of energy, to keep energy costs low enough to meet the needs of a growing economy, and to protect the environment while producing and consuming that energy. A continuing theme during this period has been that dependence on imported oil for a large share of the U.S. energy mix, particularly in the transportation sector, impedes that aim in all three dimensions. But the importance given to import dependence varies. For some, import dependence is the primary concern; for others, particularly those focused on environmental issues, it is a symptom of a general crisis that arises from indiscriminate consumption of fossil fuels. A particularly controversial aspect of the debate is the issue of global climate change, because burning fossil fuels produces large amounts of carbon dioxide, a greenhouse gas.

Like the goals of energy policy, the means of achieving them have three dimensions: reducing consumption by increased energy efficiency; increasing domestic production of conventional energy sources, particularly oil; and developing new sources of energy, particularly renewable energy and renewable fuels, that can replace oil and other fossil fuels.

Pursuing the goals of energy policy has been complicated by the diversity of energy consumption and supply in the United States. On the consumption side, there are three major sectors: residential/commercial, industrial, and transportation. On the supply side, the primary sources have traditionally been fossil energy: petroleum, natural gas (and "natural gas liquids" such as propane and butane), and coal. Electricity, which is both an energy source and a consumer of energy, has replaced some fossil fuels: about 75% of the energy consumed by the residential/commercial sector is electricity, and industrial energy consumption is about 35% electricity. But in the transportation sector, petroleum has remained dominant. Only in the past few years has corn-derived ethanol become a significant transportation fuel, replacing around 10% of gasoline consumption.

A diverse spectrum of generating sources is used to produce electricity. Coal for many years supplied half the electricity generated nationally. In recent years its share has declined; it was about 40% in December 2013. Generation by natural gas has risen in importance, supplying about 26% in December 2013. To those who regard global climate change as an urgent issue, this trend is important because generating electricity from coal emits roughly twice the carbon dioxide per kilowatt-hour than generating from natural gas. Nuclear fission supplies about 20%, hydropower less than 10%. Petroleum, an important generating fuel in the 1970s and early 1980s, now contributes less than 1% of electricity generation. A surge of construction of wind-powered generating capacity has brought its share of total generation to almost 5%.

An issue that cuts across all these factors is the role of government. How much does government policy affect energy markets? A fundamental dichotomy that lies beneath many individual policy debates, not only in energy issues, is between those who see government intervention as a positive force, and those who view it at best as a necessary evil to be restricted as much as possible.

POLICY GOALS

Conservation and Energy Efficiency

Reducing energy consumption by conservation and by increased efficiency of energy use has been a major component of policy since the first energy crisis in the 1970s. Most prominent has been setting fuel economy standards for automobiles and trucks.[1] Federal research and development programs in energy-efficient technologies have had continued support over many years, pursuing improvements in building technology, in industrial processes, and in vehicle efficiency—the last including battery designs that have been used in producing hybrid vehicles.[2] Development of "Smart Grid" technologies in electric power distribution systems to encourage more efficient use of electricity has received much recent attention. Grant programs to improve and "weatherize" existing residences have received continued funding, but like all such programs they have become controversial in the current tight budget environment.[3] Standards for home appliances such as air conditioners, refrigerators, and washing machines are another policy program that has had continued support, and some controversy. A requirement to raise efficiency levels in light bulbs has been an issue for several years.

Increasing Domestic Supply Production of Oil

With dependence on oil imports probably the most high-profile energy policy issue, the question of domestic oil production has a long history of controversy. Much exploration and development of new oil resources involves federal land, particularly on the outer continental shelf (OCS) and in Alaska, and environmental concerns have led to extended moratoria on leasing for many areas.[4] The Deepwater Horizon oil spill in the Gulf of Mexico in 2010 added a further dimension to the question of OCS leasing.[5] In recent years, development of Canadian oil sands resources, and of tight oil deposits in the United States, has added further controversy over the possible environmental effects of their production and transportation.[6] The proposed Keystone XL pipeline to bring Canadian crude to Texas refineries has been particularly controversial.[7]

In addition, for some, increasing production of oil is in principle to be avoided, since it merely postpones the replacement of fossil fuels by renewable energy and makes the transition more difficult. Others are particularly opposed to development of oil sands, on the grounds that it produces more greenhouse gases than conventional oil production.[8]

The Price of Oil and Gasoline

Since 2004, oil and gasoline markets have been highly volatile. Throughout this period some have argued that prices were being driven not by the cost of producing the resources but by speculation and unregulated manipulation of the markets. The Dodd-Frank Wall Street Reform Act of 2010 (P.L. 111-203) aimed to address these concerns, but its application and enforcement remain controversial.[9] The issue is complicated because oil prices are largely determined in a world market beyond the reach of domestic regulation.

An additional issue involving oil and gasoline prices is the role of the Strategic Petroleum Reserve (SPR), which was set up after the Arab oil embargoes to address temporary interruptions in the supply of oil. In principle releases from the SPR are limited to cases in which a physical lack of supply exists, but some have argued that it can be used to dampen surges in world oil prices even when current supply is adequate to meet demand. The June 2011 release of 30 million barrels from the SPR in response to the Libyan civil war has been deemed by some critics as such an attempt to influence the market when U.S. supplies were adequate.[10]

Natural Gas

Unlike the world oil market, in which events abroad quickly affect prices locally, the natural gas market is largely domestic. Except for about 5% net imports from Canada by pipeline, 2012 gas consumption, 25.5 trillion cubic feet, was almost entirely produced in the United States. Production since 2010 has increased sharply, largely as a result of development of tight shale formations. As with tight oil production, shale gas development has brought environmental concerns about the effects on ground water of hydraulic fracturing and horizontal drilling. The need for gathering infrastructure and pipeline construction from new fields is also an issue.

With gas production up and demand not growing as quickly, the likelihood of increasing exports of liquefied natural gas (LNG) has increased. LNG requires complex and expensive processing and loading facilities that have been controversial in locations where they have been proposed. There is also the prospect of controversy over the question of exporting energy resources such as natural gas, and its effect on prices and supply.[11]

Electric Power Production

The electric power sector since the mid-1980s has been essentially independent of the major energy policy issue of dependence on imported oil, since oil-generated electricity is a very small part of the generation mix. Nevertheless, electricity's central role in America's energy mix makes it subject to numerous controversies.

A major consideration is the role of the coal industry. More than 1 billion tons of coal are consumed in the United States each year, about 93% of it burned to produce electricity. Initiatives by the Environmental Protection Agency (EPA) that would impose tighter emissions restrictions on coal-fired power plants are particularly controversial. Limits on cross-state emissions of sulfur dioxide and nitrogen oxides, emissions of mercury and other hazardous pollutants, and regulation of greenhouse gas emissions, among other proposed regulations, have been characterized by critics as a regulatory "train wreck" that would impose excessive costs and lead to plant retirements that could threaten the adequacy of electricity capacity (i.e., reliability of supply) across the country, although some in the electric power industry consider those concerns overstated.[12]

Nuclear power is also a continuing issue. Although subject to continuing opposition over questions of safety, disposal of radioactive waste, and possible proliferation of nuclear weapons, nuclear fission has gained support because it does not emit greenhouse gases. However, cost considerations in the face of increasing natural gas production, and safety concerns enhanced by the tsunami-caused accident at Japan's Fukushima nuclear plant in March 2011, have put further expansion of nuclear power in question.[13]

Replacing Conventional Energy Sources

The third path toward reaching the goals of energy policy is to develop alternative sources of energy to replace fossil fuels. As noted, reducing the need for imported oil has been a major feature of energy policy, and congressional mandates have led to increased consumption of ethanol and other biofuels. However, essentially all fuel ethanol currently is produced from corn, potentially putting pressure on food production and food prices. The technology for producing ethanol from non-food sources (including cellulosic biomass) faces serious technological barriers. Another transportation alternative, long considered but only slowly adopted, has been natural gas-powered vehicles. Recent increases in natural gas production, noted above, have made this option appear more attractive, although developing a supply infrastructure and overcoming technological and cost difficulties continue to present barriers to widespread adoption.

For many participants in the energy debate, replacing fossil fuels has been a goal not limited to the transportation sector, however. Electricity production by renewable energy sources—wind power, concentrating solar power, photovoltaic cells, geothermal energy, and biomass—is the goal of many initiatives: research and development programs, tax benefits, loan guarantees, and mandates.

The main stimulus for these programs is environmental, including concern about global climate change, but the prospect of developing new industrial production from expansion of renewable energy sources has been a significant, if controversial, theme. Even nuclear power, long a target of environmental concerns, has been considered a viable alternative to fossil energy for electricity generation, although once again the influx of newly developed natural gas resources has raised serious questions about the cost competitiveness of new nuclear power generators.

ENERGY LEGISLATION

Earlier Legislation

Energy policy historically has been legislated mostly in large, complex bills that deal with a wide variety of issues, with debate spanning several sessions. The Energy Policy Act of 2005 (EPAct 2005; P.L. 109-58) was the most recent comprehensive general legislation, with provisions and authorizations in almost all areas of energy policy. EPAct 2005 also set up in DOE the program of energy project loan guarantees which became a source of controversy and debate following the bankruptcy of the Solyndra solar system manufacturing facility in 2011.[14]

The Energy Independence and Security Act of 2007 (EISA, P.L. 110-140) set new target fuel economy standards for cars and light trucks of 35 miles per gallon by 2020, and expanded the renewable fuels standard (RFS) to require 9.0 billion gallons in 2008 and rise to 36 billion gallons by 2022. EISA also included new efficiency standards for appliances and for light bulbs, the latter being particularly controversial in the 112th Congress.

In the 111th Congress the American Recovery and Reinvestment Act (the "Stimulus" Act, ARRA, P.L. 111-5) had major energy policy provisions, including expansion of the loan guarantee program and large increases in funding for renewable energy programs. The Department of Energy's Office of Energy Efficiency and Renewable Energy, in addition to the $2 billion appropriated in the FY2009 regular appropriations bill, received $17 billion in ARRA, of which $11.5 billion was for grants to states for energy, efficiency, and weatherization programs. The Office of Electricity Delivery and Energy Reliability, which had historically been funded at about $150 million per year, received $4.5 billion in ARRA, directed at establishing "Smart Grid" technology for the electric power industry.

Legislation in the 113th Congress

The 113th Congress has not dealt with major energy legislation, but numerous energy policy questions have been taken up in proposed legislation. Congress has enacted statutes on generally less controversial topics, including hydroelectric licensing and permitting and pipeline safety documentation (P.L. 113-30). The 113th Congress also enacted the Water Resources Reform and Development Act of 2014 (WRRDA, P.L. 113-121), omnibus reauthorization of water resource activities.

More controversial energy policy topics have seen committee and floor action since the start of the 113th Congress, including legislation on the Keystone XL pipeline, hydraulic fracturing, and energy efficiency. Some of these proposals have been rolled into H.R. 2, which was introduced September 15, 2014, and considered by the House Committee on Rules the following day. Floor debate on the bill was scheduled for September 18.

Keystone XL Pipeline and Other Pipeline Legislation

In the ongoing debate over approval of the proposed Keystone XL pipeline to bring Canadian oil sands crude oil to the Gulf of Mexico, H.R. 3, the Northern Route Approval Act, would declare that a presidential permit would not be required for its construction.[15] The bill passed the House on May 22, 2013, by a vote of 241-175.

In response to the wide spread of new production of natural gas and a consequent pressure to expand the means of transporting it to market, the House passed legislation (H.R. 1900) on November 21, 2013, to require the Federal Energy Regulatory Administration (FERC) to respond to new gas pipeline permit applications within one year.[16]

Increasing Domestic Production

A bill to increase on-shore leasing of oil and gas resources on federal land (H.R. 1965), including the National Petroleum Reserve in Alaska, and shale oil deposits in Colorado, passed the House on November 21, 2013. A bill promoting increased offshore oil and gas exploration (H.R. 2231) passed the House on June 28, 2013.

Regulation of Power Plant Emissions and Global Climate Change

As noted above, EPA's moves to regulate emissions from coal-fired power plants have been controversial, and several bills in the 112th Congress passed the House without being taken up in the Senate. Similar action has taken place in the 113th Congress.

On June 25, 2013, President Obama announced a national plan to reduce emissions of carbon dioxide and other greenhouse gases, as well as to encourage adaptation to expected climate change. A major feature of the President's Climate Action Plan was a directive to the Environmental Protection Agency to issue rules to curtail carbon dioxide emissions from new and existing power plants.[17] On March 6, 2014, the House passed H.R. 3826, a bill to block the Environmental Protection Agency from limiting power plant carbon emissions, by a vote of 229- 183.

Hydraulic Fracturing

The issue of regulation and safety of hydraulic fracturing led to differing legislative proposals. Under provisions of EPAct 2005, EPA does not have regulatory jurisdiction over certain aspects of hydraulic fracturing. In the 112th Congress, one proposed legislative solution, the Fracturing Responsibility and Awareness of Chemicals Act (FRAC Act, H.R. 1084 and S. 587), would have given EPA such jurisdiction under the Safe Drinking Water Act. Another bill would have maintained jurisdiction within the states. Neither legislative proposal was reported out of committee. In the 113th Congress, the FRAC Act was reintroduced as H.R. 1921.

Another aspect of the hydraulic fracturing issue is the proposal by the Bureau of Land Management (BLM) to regulate fracking on federal lands.[18] On November 20, 2013, the House passed H.R. 2728, which would prohibit the Department of the Interior from enforcing rules related to hydraulic fracturing in "any state that has regulations, guidance or permit requirements for that activity."

Energy Efficiency

The Energy Savings and Industrial Competitiveness Act of 2013, S. 2262 and H.R. 1616, would promote energy efficiency in buildings and industry by encouraging adoption of uniform building codes and authorize a grant program for state energy efficiency programs. Several versions of the bill have seen committee and/or floor action in the Senate, but S. 2262 failed a cloture vote on May 12, 2014. On March 5 the House passed H.R. 2126, which contains some of the provisions in S. 2262, by a vote of 375-36.

Multi-Faceted Legislation

On September 16, 2014, the House Rules Committee approved a closed rule for H.R. 2, which contains provisions from several bills, including H.R. 3, H.R. 1900, H.R. 1965, H.R. 2231, and H.R. 3826. The bill would, among other provisions, effectively approve the Keystone XL pipeline, amend FERC authority to approve natural gas pipelines, increase federal on-shore and off-shore oil and gas leasing, and limit EPA's authority to regulate greenhouse gas emissions. Floor debate on the bill was scheduled for September 18.

End Notes

[1] For details, see CRS Report R42721, Automobile and Truck Fuel Economy (CAFE) and Greenhouse Gas Standards, by Brent D. Yacobucci, Bill Canis, and Richard K. Lattanzio.

[2] See CRS Report R43567, Energy and Water Development: FY2015 Appropriations, coordinated by Mark Holt.

[3] See CRS Report R42147, DOE Weatherization Program: A Review of Funding, Performance, and Cost-Effectiveness Studies, by Fred Sissine.

[4] See CRS Report R42432, U.S. Crude Oil and Natural Gas Production in Federal and Non-Federal Areas, by Marc Humphries, and CRS Report R41132, Outer Continental Shelf Moratoria on Oil and Gas Development, by Curry L. Hagerty.

[5] See CRS Report R41684, Enacted and Proposed Oil Spill Legislation in the 112th Congress, by Jonathan L. Ramseur.

[6] See CRS Report R41760, Hydraulic Fracturing and Safe Drinking Water Act Regulatory Issues, by Mary Tiemann and Adam Vann.

[7] See CRS Report R41668, Keystone XL Pipeline Project: Key Issues, by Paul W. Parfomak et al., and CRS Report R42611, Oil Sands and the Keystone XL Pipeline: Background and Selected Environmental Issues, coordinated by Jonathan L. Ramseur.

[8] See CRS Report R42537, Canadian Oil Sands: Life-Cycle Assessments of Greenhouse Gas Emissions, by Richard K. Lattanzio.

[9] See CRS Report R41398, The Dodd-Frank Wall Street Reform and Consumer Protection Act: Title VII, Derivatives, by Rena S. Miller and Kathleen Ann Ruane.

[10] See CRS Report R42460, The Strategic Petroleum Reserve: Authorization, Operation, and Drawdown Policy, by Anthony Andrews and Robert Pirog.

[11] See CRS Report R42074, U.S. Natural Gas Exports: New Opportunities, Uncertain Outcomes, by Michael Ratner et al.

[12] See CRS Report R41914, EPA's Regulation of Coal-Fired Power: Is a "Train Wreck" Coming?, by James E. McCarthy and Claudia Copeland.

[13] See CRS Report RL33558, Nuclear Energy Policy, by Mark Holt.

[14] See CRS Report R42152, Loan Guarantees for Clean Energy Technologies: Goals, Concerns, and Policy Options, by Phillip Brown.

[15] For more information, see CRS Report R41668, Keystone XL Pipeline Project: Key Issues, by Paul W. Parfomak et al.

[16] For more information, see CRS Report R43138, Interstate Natural Gas Pipelines: Process and Timing of FERC Permit Application Review, by Paul W. Parfomak.

[17] See CRS Report R43120, President Obama's Climate Action Plan, coordinated by Jane A. Leggett.

[18] See CRS Report R43148, An Overview of Unconventional Oil and Natural Gas: Resources and Federal Actions, by Michael Ratner and Mary Tiemann.

INDEX

A

B

C

D

E

F

G

H

I

O

P

T

U

V

W

Y